Biology
Miller & Levine

Probeware Lab Manual
for Computers and Calculators

Boston, Massachusetts
Chandler, Arizona
Glenview, Illinois
Upper Saddle River, New Jersey

13-digit ISBN 978-0-13-369006-4
10-digit ISBN 0-13-369006-7
4 5 6 7 8 9 10   V016   14 13 12 11

Contents

To The Teacher

The *Miller & Levine Biology Probeware Lab Manual* contains student versions of labs for use with computers, handheld interfaces, and calculators. (Annotated versions of the labs are available on the accompanying CD-ROM.) All the labs in this manual have been created to directly support your curriculum goals. Initially, you may find that students progress slowly, but as they become familiar with the software and hardware, the benefits of doing probeware labs will quickly become evident. Probeware sensors and user-friendly software simplify data collection in the classroom and in the field. As students become familiar with the probeware software and hardware, collecting, graphing, and analyzing data become almost effortless. By simplifying the data-gathering process, probeware lets students focus on drawing conclusions. Students then reap the rewards of a much fuller conceptual understanding.

The *Miller & Levine Biology Probeware Lab Manual* is designed to meet your needs whether you are using PASCO®, Vernier, or Texas Instruments hardware and software. The labs in the manual have been developed in correlation to specific content in the *Miller & Levine Biology* student edition. They also follow the same structure as other labs within the *Miller & Levine Biology* program. The printed workbook contains the student editions of 15 probeware labs. There are five labs for each specific hardware and software provider.

On the following pages, you will find information about each provider of hardware and software. Additional information can be found in user guides included on the CD-ROM, with your probeware equipment, or at the provider's Web site.

The CD-ROM includes both editable and PDF versions of the student edition of each lab, as well a PDF teaching guide for each lab. The CD-ROM also includes technical tips for specific use of equipment provided by the hardware and software companies.

As always, safety must be the highest priority in the laboratory. Please make sure you and your students have reviewed the safety instructions and guidelines on pages xiii–xvi.

Correlation to
Miller & Levine Biology Student Edition

PASCO

Lab	Biology Chapter
1. Enzyme Action	Chapter 2 The Chemistry of Life
2. Microclimates	Chapter 4 Ecosystems and Communities
3. Respiration in Seeds	Chapter 9 Cellular Respiration and Fermentation
4. Transpiration	Chapter 23 Plant Structure and Function
5. Regulation of Body Heat	Chapter 30 Digestive and Excretory Systems

VERNIER

Lab	Biology Chapter
6. Enzyme Action	Chapter 2 The Chemistry of Life
7. Testing Watersheds	Chapter 4 Ecosystems and Communities
8. Comparing Respiration Rates in Seeds	Chapter 9 Cellular Respiration and Fermentation
9. Transpiration Rates	Chapter 23 Plant Structure and Function
10. Human Respiration	Chapter 33 Circulatory and Respiratory Systems

TEXAS INSTRUMENTS

Lab	Biology Chapter
11. Enzyme Action	Chapter 2 The Chemistry of Life
12. Testing Watersheds	Chapter 4 Ecosystems and Communities
13. Comparing Respiration Rates in Seeds	Chapter 9 Cellular Respiration and Fermentation
14. Transpiration Rates	Chapter 23 Plant Structure and Function
15. Human Respiration	Chapter 33 Circulatory and Respiratory Systems

PASCO® Scientific

21ˢᵗ Century Science Education Solutions

Every 21st century science teacher needs access to devices that enable them to emulate modern 21st century practices in science. PASCO's PASPORT® line of data-collection systems offers a flexible and affordable approach to teaching science with technology.

In the *Probeware Lab Manual,* the term *data-collection system* refers to the data-collection, display, and analysis device used to carry out the five biology lab activities from PASCO, customized for *Miller & Levine Biology.* These activities include computer-based software (SPARKvue™ and DataStudio®) and computer-free devices (SPARK Science Learning System™ and Xplorer GLX®). For details on each of the technologies, visit www.pasco.com/products.

Students use powerful onboard graphing and analysis tools to analyze collected data. The activities are designed so that any PASCO data-collection system can be used to carry out the procedure. Tech Tips provide the steps on how to use a specific data-collection system.

Biology Lab Activities and Probeware

PASCO's biology labs in the *Probeware Lab Manual* follow the same inquiry-based format as defined by *Miller & Levine Biology* throughout the textbook. (See www.pasco.com for the *Miller & Levine Biology* Bundle.)

Accompanying PASCO Materials

In addition to the five PASCO biology lab activities customized for *Miller & Levine Biology,* the CD-ROM contains the following materials from PASCO:

- Tech Tips for using each data-collection system and sensor technologies.
- Four Scientific Inquiry starter labs, one for each data-collection system. The teacher and student versions of the starter labs are in PDF format.
- Quick Start Guides for SPARK Science Learning System, Xplorer GLX, and DataStudio in PDF format.

Guideline to Implementing Labs in the Classroom

PASCO provides a variety of materials to help teachers start using sensors in the science classroom. Teachers unfamiliar with the use of sensors should start with the tutorials and follow the suggested sequence outlined below. Teachers already familiar with PASCO sensor use in their classrooms can launch right into using the labs but should familiarize themselves with the Tech Tips.

Getting Started Tutorials

To help teachers and students become familiar with the many features of their data-collection system, tutorials are available for the SPARK Science Learning System, SPARKvue, Xplorer GLX, and DataStudio on the CD-ROM that comes with the equipment. Alternatively, they can be viewed on PASCO's Web site, www.pasco.com/resources/.

Quick Start Guides

Teachers and students can get a quick overview of the features of their data-collection system from the Quick Start Guides available on the CD-ROM for the SPARK Science

Learning System, Xplorer GLX, and DataStudio. There is no Quick Start Guide for SPARKvue as the features and functionality of SPARKvue and the software on the SPARK Science Learning System are almost identical. Complete user's guides for each of the platforms are available on PASCO's Web site, www.pasco.com/downloads/.

Starter Labs

The Scientific Inquiry activity, available on the CD-ROM in the Starter Lab folder, is the first activity students should perform. It serves two purposes:

1. It introduces students to scientific inquiry, that is, the process of conducting science investigations (the scientific method).
2. Each data-collection system has its own custom Scientific Inquiry activity so that teachers and students are introduced easily and quickly to the commonly used features of their device.

Start with the activity specific to your data-collection system to become familiar with the data-collection system.

Lab Procedure and Tech Tips

The Procedure section directs the hands-on portion of the lab activity. Each step that requires a technical procedure using the data-collection system has associated instructions (referred to as "Tech Tips") available on the CD. Students who need technical help to complete the process can consult the Tech Tips that contains the instructions for their particular data-collection system. Excerpts from one of the lab procedures and from the associated Tech Tips illustrate how to use the Tech Tips while following the procedures.

Note the reference to the Tech Tip at the end of a procedure (e.g., $^{1.2}$). Refer to the Tech Tip in the PDF for the specific data-collection system being used. These instructions are for SPARKvue.

Steps in Lab Procedure With Tech Tip Identification	Associated Tech Tip Instructions for Each Procedure
2. Using a sensor extension cable, connect a barometer/low pressure sensor to the data-collection system.$^{2.1}$	**2.1 Adding a sensor to your SPARKvue experiment** 1. Connect an interface such as a SPARKlink or a SPARK Science Learning System to your computer. 2. Connect a sensor to the interface. *Task result*: SPARKvue detects the sensor.

Provide a copy of the Tech Tips to each group of students, along with the user guides for the sensors used in the labs.

Technical and Teacher Support

For information about or help with PASCO products, contact PASCO's Technical and Teacher Support group by phone, e-mail, or on the Web.

Phone: 1-800-772-8700 (in the U.S.)
+1-916-786-3800 (worldwide)

E-mail: support@pasco.com

Web: www.pasco.com/support

Vernier
Hardware and Software Information

Vernier Software & Technology has been an innovator of data-collection technology for more than 25 years. Vernier products play a wide-ranging role in today's science education, from elementary-school activities to college-level courses. By using Vernier's easy-to-use and affordable science interfaces, sensors, and software, teachers can enhance their science curriculum and build students' critical thinking skills.

Vernier offers data-collection solutions for computers, Texas Instruments graphing calculators, and stand-alone data collection with the Vernier LabQuest Interface. Over 50 different sensors are available from Vernier; these sensors support Auto-ID technology to provide user-friendly data collection. The award-winning Logger *Pro* data-collection software includes a generous site license policy permitting installation on all computers in a K–12 school or university department. In addition, each copy comes with a free student license allowing installation on students' home computers. A free demonstration version of Logger *Pro* can be downloaded from Vernier's Web site at www.wernier.com/downloads/lp3demo.html.

For more information on Vernier's sensors, software, training, and correlations to state standards, visit www.vernier.com.

Contact Information

For customer support, product information, and current prices, please contact

Vernier Software and Technology
13979 SW Millikan Way
Beaverton, OR 97005-2886

Phone: 1-888-837-6437
Fax: 503-277-2440
Web site: www.vernier.com
E-mail: info@vernier.com

Vernier and *Miller & Levine Biology* Labs

Vernier has customized five labs for the *Miller & Levine Biology* program. The labs have been written to work with any of the Vernier data-collection systems. The content and structure of the labs have been written in the same format as other labs for *Miller & Levine Biology*. There is a step-by-step guide provided on the CD-ROM to show students and teachers information specific to the use of their data-collection systems and sensors.

Vernier Technical Guide

For steps in the procedure that require technical instructions, students and teachers should refer to the Vernier Technical Guide PDF in the Vernier folder on the CD-ROM. Steps with specific instructions have a superscripted number that corresponds to the Vernier Technical Guide. The following is an example of how to use the guide:

Step in Procedure	From Vernier Technical Guide
14. Start data collection.[2]	**2.** Start Data Collection **a.** If using Computer and Logger*Pro:* Click `▶ Collect` to begin data collection. **b.** If using LabQuest®: Click to begin data collection.

Getting Started With Logger *Pro*®

Before using the labs in this probeware manual, you must first install the most recent version of Logger *Pro* software on your computer. Logger *Pro* software is sold separately by Vernier Software & Technology. If you do not have a copy of this software, a demonstration version is available at Vernier's Web site at www.vernier.com/downloads/lp3demo.html.

Follow these simple steps to collect data:

- Connect the LabPro® or LabQuest™ interface to the computer.
- Connect the desired sensor to the interface.
- Launch the Logger *Pro* software.
- Open the experiment file indicated in the lab.
- Click `▶ Collect` to begin data collection.

Texas Instruments Technology Information

Let your students discover science in the world around them by helping them develop hypotheses, complete experiments, and form conclusions.

The CBL 2™

The CBL 2™ allows students to collect and analyze real-world data using Texas Instruments graphing calculators.

You can collect data with the Verier EasyData® Application and one of more than 50 available sensors from Vernier Software & Technology. The range of sensors available for the CBL 2 allows you and your students to conduct a wide variety of biology investigations. More details about the CBL 2 features are available in the CBL 2 Technical Reference found on the CD-ROM in this lab manual. For additional information on Texas Instruments products, visit the TI Web site at education.ti.com/datacollection.

General Texas Instruments Information

Contact the TI Customer Support Line to order products, before returning a product for service, or if you have general questions about using a product.

Call: 1-800-TI-CARES (800-842-2737)
Monday–Thursday: 8:00 A.M. to 7:00 PM Central Time
Friday: 10:00 AM to 7:00 PM Central Time

E-mail: ti-cares@ti.com

Write: Texas Instruments
Customer Support Line
PO Box 650311, MS 3962
Dallas, TX 75265

Technical Assistance

The TI Technical Assistance Group is trained to answer your technical questions about TI calculators, software, and accessories.

Call: 972-917-8324 (This is not a toll-free number.)
Monday–Thursday: 8:00 A.M. to 4:30 PM Central Time
Friday: 10:00 AM to 4:30 PM Central Time

E-mail: ti-cares@ti.com

Write: Texas Instruments
Technical Assistance Group
PO Box 650311, MS 3962
Dallas, TX 75265

Getting Started With EasyData®

Vernier's EasyData Application for TI-83/84 Plus family of graphing calculators is used to collect, graph, examine, and analyze data. Follow these simple steps to collect data:

- Connect the CBL 2™ or LabPro® interface to the calculator.
- Connect the desired sensor to the interface.
- Press ⌈APPS⌉ and scroll to the EasyData App.
- Press ⌈ENTER⌉ to launch EasyData.
- Select ⌈START⌉ to begin data collection.

Capturing a Calculator Graph for Printing

Many labs in the probeware manual contain steps to create a graph. You may wish to have your students print the graph directly from the calculator. To capture and print a graph, you will need a TI Connectivity cable and the TI Connect Software. Before doing an experiment that requires printing, read the instructions listed below.

Windows Computers

1. On your calculator, display the graph you want to print.
2. Connect the TI Connectivity cable to your computer and TI graphing calculator.
3. Start the TI Connect software on your computer.
4. Within Connect, click on Screen Capture. The captured screen will appear in the TI Screen Capture window.
5. Click on the print icon to print the graph.

Macintosh Computers

1. On your calculator, display the graph you want to print.
2. Connect the TI Connectivity cable to your computer and TI graphing calculator.
3. Start the TI Connect software on your computer.
4. Double-click on the TI Device Explorer. The TI Device Explorer window will appear, displaying the contents of the calculator's memory.
5. Click the "Screen Capture" toolbar control. The calculator's screen image will appear in a separate window.
6. Click on the print icon to print the graph.

Safety in the Biology Laboratory

Working in the biology laboratory can be exciting, but it can also be dangerous if you do not follow the proper safety guidelines. You are responsible for helping to maintain a safe environment in the laboratory. Unsafe practices endanger not only you but also the people who work near you.

Safety Rules

To prepare yourself for a safe year in the laboratory, review the following safety rules. Make sure you understand each rule. Ask your teacher to explain any rules you do not understand.

Dress Code

- To protect your eyes, wear safety goggles when you see the safety goggles symbol. This symbol will appear when you work with chemicals that are corrosive, such as acids or bases. The symbol will also appear when a chemical has irritating vapors.

  Avoid wearing contact lenses when you work with chemicals. If you need to wear contact lenses in the lab to see clearly, talk with your teacher about your need.

- Wear a laboratory apron or coat when you are working with chemicals that are corrosive or can stain your clothing. Also wear an apron when you are heating materials.

- Tie back long hair to avoid contact with chemicals, flames, or biological cultures.

- Remove, tie back, or avoid wearing any clothing or jewelry that can hang down and touch chemicals, flames, or cultures.

- Do not wear sandals or open-toed shoes in the laboratory. Never walk around barefoot or in stocking feet in the laboratory.

General Safety Rules

- Read the procedure for a lab in advance. Follow the steps of the procedure exactly as they are written unless your teacher tells you otherwise. If you do not understand a step, ask your teacher for help.

- Never do an experiment that your teacher has not approved. Do not use any lab equipment without permission or without the supervision of your teacher.

- Never eat or drink anything while you are in the laboratory. Do not bring food into the lab that you intend to eat later. Do not chew gum. Do not apply cosmetics.

- If you spill a chemical, check with your teacher right away about the proper cleanup procedure.

- Do not pour chemicals or other materials into the sink or place items in a trash container unless specifically instructed to do so by your teacher.

Emergencies and First Aid

- Know the location and proper use of safety equipment, such as the fire extinguisher, fire blanket, eye-wash station, and first-aid kit.

- Learn what to do in response to specific emergencies, such as cuts, burns, or contact with chemicals.

- Immediately report all accidents, no matter how minor, to your teacher.

- Your teacher will help you determine the proper response to an accident. He or she may administer first aid, send you to the school nurse, or call a physician.

- Report any fires to your teacher at once. Know the location of the fire alarm, and know where and how to report a fire or other emergency requiring outside assistance.

Heating and Fire Safety

- Always wear safety goggles when you use a candle or gas burner as a heat source.
- Make sure you know how to safely light a gas burner. Your teacher will demonstrate the proper procedure for lighting a burner.
- If the flame leaps out of a burner toward you, turn the gas off immediately. Do not touch the burner, which may be hot.
- Never leave a lighted burner unattended.
- Never reach across a flame.
- Never heat a chemical that you are not told to heat. A chemical that is harmless when cool can be dangerous when heated.
- Make sure that there are no open containers of flammable liquids in the laboratory when flames are being used.
- When you heat a test tube with a flame, point the opening of the tube away from yourself and others. Chemicals can splash or boil out of a heated test tube.
- Never heat a closed container. Expanding gases inside the container may cause the container to blow apart, causing injury to you or others working nearby.
- Never pick up a container that has been heated without first holding the back of your hand near it. If you can feel the heat on the back of your hand, the container may be too hot to handle. Use a clamp or tongs when handling hot containers, or, if appropriate, wear heat-resistant gloves.
- Maintain a clean work area and keep materials other than the ones you are heating away from flames or hot plates.

Using Chemicals Safely

- Never mix chemicals for "the fun of it." You might produce a dangerous, possibly explosive, substance.
- Chemicals used in labs may be poisonous. Never touch, taste, or smell a chemical that you do not know for a fact is harmless. If you are instructed to smell the fumes in an experiment, gently wave your hand over the opening of a container and direct the fumes toward your nose. Do not inhale the fumes directly from the container.
- Use only those chemicals needed in the investigation. Keep all lids closed when a chemical is not being used. Notify your teacher whenever chemicals are spilled.
- Dispose of all chemicals as instructed by your teacher. To avoid contamination, never return chemicals to their original containers.
- Be extra careful when working with acids or bases. Pour such chemicals from one container to another over the sink, not over your work area.
- When diluting an acid, pour the acid into water. Never pour water into the acid.
- Notify your teacher immediately if you spill an acid or a base.
- Use a continuous stream of water from the tap to rinse acid or bases off your skin or clothing.
- If an acid or base splashes in your eyes, go to the eye-wash station immediately to flush the liquid from your eyes.

Using Glassware Safely

- Before you heat a beaker or other glass container, make sure the outside of the container is dry.
- If you use a gas burner as a heat source, use a wire screen to protect glassware from the flame.
- Keep in mind that hot glassware will not appear hot. Never pick up a heated glass object without first checking to see if it is hot. If you can feel the heat on the back of your hand, use a clamp or tongs to handle the container, or wear heat-resistant gloves if appropriate.
- Never use broken or chipped glassware.
- If a glass object breaks, notify your teacher. Then use a brush and dustpan to collect and dispose of the broken glass in the proper trash container. Never pick up broken glass with your hands.
- Never eat or drink from laboratory glassware. Thoroughly clean glassware before putting it away.
- If you need to insert a glass tube into a rubber stopper or rubber tubing, use a lubricant and a gentle turning motion to reduce the risk of breaking the glass.

Using Sharp Instruments

- Handle scalpels, scissors, and knives with extreme care. Always cut away from yourself.
- Always use sharp instruments only as instructed.
- Notify your teacher immediately if you cut yourself while working in the laboratory.

Working With Live Organisms

- No experiments that will cause pain, discomfort, or harm to animals should be done in the classroom or at home.
- Treat all living organisms with care and respect. Limit your interaction with organisms to only what is required to complete an experiment.
- Pay close attention to the instructions in the procedure and any additional instructions from your teacher.
- Use sterile procedures, as instructed by your teacher, when working with microbes.
- Wash your hands thoroughly after you handle any organisms or their containers.

End-of-Experiment Rules

- Before you leave the laboratory, clean up your work area and return all equipment to its proper place.
- Make sure you turn off and disconnect burners from the gas supply. Also turn off and unplug hot plates.
- Wash your hands thoroughly with soap and warm water before you leave the lab.

Safety Alerts

All the labs in this manual have been designed with safety in mind. When appropriate, you will find a paragraph about safety after the list of materials. Some of the safety symbols shown on the next page will also be included to alert you to possible dangers and to remind you to work carefully.

Watch for statements within a procedure that are labeled "CAUTION." These statements focus on possible risks related to specific steps in the procedure. The possible risk may be to yourself, to an organism, or to equipment.

Safety Symbols

Safety Goggles Always wear safety goggles to protect your eyes during any activity involving chemicals, flames or heating, or the possibility of flying objects, particles, or substances.

Lab Apron Wear a laboratory apron to protect your skin and clothing from injury.

Plastic Gloves Wear disposable plastic gloves to protect yourself from contact with chemicals or organisms that could be harmful. Keep your hands away from your face, and dispose of the gloves according to your teacher's instructions at the end of the activity.

Breakage Handle breakable materials such as thermometers and glassware with care. Do not touch broken glass.

Heat-Resistant Gloves Use an oven mitt or other hand protection when handling hot materials. Heating plates, hot water, and glassware can cause burns. Never touch hot objects with your bare hands.

Heating Use a clamp or tongs to hold hot objects. Do not touch hot objects with your bare hands.

Sharp Object Scissors, scalpels, pins, and knives are sharp. They can cut or puncture your skin. Always direct sharp edges and points away from yourself and others. Use sharp instruments only as directed.

Electric Shock Avoid the possibility of electric shock. Never use electrical equipment around water or when the equipment or your hands are wet. Be sure cords are untangled and cannot trip anyone. Disconnect equipment when it is not in use.

Flames Tie back loose hair and clothing, and put on safety goggles before working with fire. Follow instructions from your teacher about lighting and extinguishing flames.

No Flames Flammable materials may be present. Make sure there are no flames, sparks, or exposed sources of heat present.

Corrosive Chemical This symbol indicates the presence of an acid or other corrosive chemical. Avoid getting the chemical on your skin or clothing, or in your eyes. Do not inhale the vapors. Wash your hands when you are finished with the activity.

Poison Do not let any poisonous chemical get on your skin, and do not inhale its vapor. Wash your hands when you are finished with the activity.

Fumes Poisonous or unpleasant vapors may be produced. Work in a ventilated area or, if available, in a fume hood. Avoid inhaling a vapor directly. Test an odor only when directed to do so by your teacher, using a wafting motion to direct the vapor toward your nose.

Physical Safety This activity involves physical movement. Use caution to avoid injuring yourself or others. Follow instructions from your teacher. Alert your teacher if there is any reason that you should not participate in the activity.

Animal Safety Treat live animals with care to avoid injuring the animals or yourself. Working with animal parts or preserved animals may also require caution. Wash your hands when you are finished with the activity.

Plant Safety Handle plants only as your teacher directs. If you are allergic to any plants used in an activity, tell your teacher before the activity begins. Avoid touching poisonous plants and plants with thorns.

Disposal Chemicals and other materials used in the activity must be disposed of safely. Follow the instructions from your teacher.

Hand Washing Wash your hands thoroughly when finished with the activity. Use soap and warm water. Lather both sides of your hands and between your fingers. Rinse well.

General Safety Awareness You may see this symbol when none of the symbols described earlier applies. In this case, follow the specific instructions provided.

Materials List

Hardware and Software

PASCO

Item	Quantity per Group	Lab Number
Data-Collection Interface (SPARK™, Xplorer GLX®, or computer-based software SPARKvue™ and DataStudio®)	1	1, 2, 3, 4, 5
PASPORT Oxygen Gas Sensor	1	1
PASPORT Quad Temperature Sensor	1	2, 5
PASPORT Carbon Dioxide Gas Sensor	1	3
PASPORT Barometer/Low Pressure Sensor	1	4
Sensor Extension Cables	1	3, 4

VERNIER

Item	Quantity per Group	Lab Number
Data-Collection Interface (LabQuest®, LabPro®, or computer-based software Logger Pro®)	1	6, 7, 8, 9, 10
Vernier Oxygen Gas Sensor	1	6
Vernier Dissolved Oxygen Sensor and calibration bottle	1	7
Vernier pH Sensor	1	7
Vernier Conductivity Sensor	1	7
Vernier Temperature Sensor	1	7
Vernier Carbon Dioxide Gas Sensor	1	8
Vernier Gas Pressure Sensor	1	9, 10
Vernier Respiration Belt	1	10

TEXAS INSTRUMENTS

Item	Quantity per Group	Lab Number
DataMate or EasyData program	1	11, 12, 13, 14, 15
LabPro or CBL 2 Interface	1	11, 12, 13, 14, 15
TI graphing calculator	1	11, 12, 13, 14, 15
Vernier Oxygen Gas Sensor	1	11
Vernier Dissolved Oxygen Sensor and calibration bottle	1	12
Vernier pH Sensor	1	12
Vernier Conductivity Sensor	1	12
Vernier Temperature Sensor	1	12
Vernier Carbon Dioxide Gas Sensor	1	13
Vernier Gas Pressure Sensor	1	14, 15
Vernier Respiration Belt	1	15

Comsumables

Item	Quantity per Group	Lab Number
Adhesive covers or tape	2	5
Dissolved oxygen filling solution	Varies	7, 12
Distilled water	Varies	1, 3, 4, 5, 6, 7, 8, 11, 12, 13
Drinking cup	1	7, 10, 12, 15
Glycerin	1 mL	4
Hydrogen peroxide solution, 3%	Varies	1, 6
Ice	Varies	1, 3, 5, 6, 8, 11, 13
Paper towel	1	8, 13
Permanent marker	1	1, 6, 7, 11, 12
Petroleum jelly	2–3 g	4
Plant seedlings (tomatoes)	1	4
Plastic trash bag, 40 L	1	10, 15
Seeds (beans or peas)	30–50	3, 8, 13
Yeast suspension	30 mL	1, 6, 11

Nonconsumables

Item	Quantity per Group	Lab Number
Beaker, 100 mL	Varies	7, 8, 12, 13
Beaker, 400 mL	1	6, 11
Beaker, 500 mL	1	1
Beaker, 1 L	2	1, 3
Bowl, large	1	4, 5
Clamp, three-fingered	1	4
Clamp, utility	1	4
Fan	1	4, 5
Glove or mitten	1	5
Graduated cylinder, 10 mL	1	6, 11
Graduated cylinder, 25 mL	1	1
Hot plate	1	1, 6, 11
Knife	Varies	5
Large base and support rod	1	4
Nalgene bottle, 250 mL	1	6, 11
Pipette or dropper	Varies	4, 6, 11
Plastic tube	1	4, 9, 14
Sampling bottle/respiration chamber	1	1, 3, 8, 13
Test tube	Varies	1, 6, 11
Test tube rack	1	6, 11
Thermometer	1	6, 8, 11, 13
Tongs	1	1
Towel	Varies	5
Water sampling bottles	4	7, 12

PASCO Labs

Teacher Notes and Answers are on the CD-ROM.

The lab activities from PASCO are listed in the table below, along with the sensor required to conduct each lab. (See www.pasco.com for the Miller & Levine Biology Bundle.)

Lab Activity	Probeware
Enzyme Action	PASPORT Oxygen Gas Sensor
Microclimates	PASPORT Quad Temperature Sensor
Respiration in Seeds	PASPORT Carbon Dioxide Gas Sensor
Transpiration	PASPORT Barometer/Low Pressure Sensor
Regulation of Body Heat	PASPORT Quad Temperature Sensor

The CD-ROM provided in the front of the book contains both a PDF and an editable copy of the student version of each lab. In addition, the CD-ROM includes a teacher's version with an answer guide and teaching tips. Finally, the CD-ROM includes PASCO's Tech Tips for technical help for your particular data-collection system.

These labs were adapted from labs created and developed by PASCO® scientific.

Lab 1 Enzyme Action

Problem

How does temperature affect the rate of an enzyme-catalyzed reaction?

Introduction

Enzymes are proteins that function as catalysts, helping to speed up reactions. Many factors affect how well enzymes are able to function. For example, all enzymes have a temperature at which they work most efficiently. What happens when the temperature falls outside that range? We will look at the enzyme catalase to find out.

Cells sometimes produce hydrogen peroxide, a toxic chemical, to help fight infections. At low levels hydrogen peroxide in a cell will fight off infection. However, at high levels hydrogen peroxide can harm a cell, so levels must be regulated. The enzyme catalase breaks down hydrogen peroxide (H_2O_2) into two harmless products: water (H_2O) and oxygen gas (O_2). In this lab, you will observe this enzyme-catalyzed reaction. You will examine the effects of temperature on the efficiency of catalase.

Skills Focus

Measure, Analyze Data, Infer, Use a Data-Collection System and Sensors

Materials

- PASCO data-collection system
- oxygen gas sensor
- sampling bottle (provided with the sensor)
- graduated cylinder, 25-mL
- beakers (2), 1-L
- beaker, 500-mL
- tongs
- catalase source: yeast suspension
- hydrogen peroxide, 3%
- ice, crushed or cubed
- test tube (2)
- water, distilled, 500 mL
- hot plate

Safety

Always wear goggles and an apron when using a water bath. If you have glass beakers or test tubes, check for cracks or chips. Alert your teacher if you break a glass object. Avoid contact with the surface of the hot plate. To avoid electric shocks, make sure cords, connections, and your hands are dry. Wash your hands thoroughly with soap and warm water at the end of the lab.

Pre-Lab Questions

1. **Relate Cause and Effect** How will you know that a chemical reaction is taking place in Part B?

   _____

   _____

2. **Predict** Do you think the heated, room-temperature, or chilled catalase will produce oxygen at a faster rate? Why?

   _____

   _____

   _____

3. **Infer** What do you think would happen if the hydrogen peroxide was not diluted with water?

   _____

   _____

   _____

Procedure

Part A: Setting Up the Experiment

1. Start a new experiment on the data-collection system. [1,2]

2. Put on goggles and an apron.

3. Pour 200 mL of water into a 1-L beaker and place the beaker on the hot plate. Heat the water until it reaches a gentle boil. **CAUTION:** Make sure your water bath is placed away from the edge of the lab bench or table.

4. Pour 10 mL of the yeast suspension into a test tube. Place the test tube in the boiling water and heat for 10 to 15 minutes as shown in Figure 1.

Figure 1 Placement of test tube in water bath.

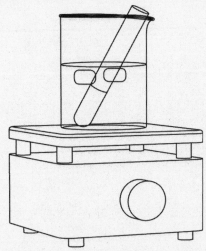

5. While that yeast suspension heats, pour 10 mL of yeast suspension into a second test tube. Place the test tube into a 1-L beaker and pack crushed or cubed ice around the test tube. Add some water and cool the test tube in the ice for 10 to 15 minutes.

6. While the yeast suspensions heat and cool, connect the cable of the oxygen-sensing element to the top of the sensing box. Connect the oxygen gas sensor to your data-collection system. [2.1]

7. Calibrate the oxygen sensor. [3.5]

8. Display Oxygen Concentration (%) on the *y*-axis versus Time on the *x*-axis of a graph. [7.1.1]

Part B: Collecting Data

9. Pour 10 mL of 3% hydrogen peroxide into the sampling bottle.

10. Add 40 mL of distilled water to the hydrogen peroxide in the sampling bottle.

11. Add 10 mL of fresh, room-temperature yeast suspension to the dilute hydrogen peroxide solution in the sampling bottle.

12. Loosely insert the oxygen-sensing element into the sampling bottle.

13. Adjust the scale of the graph to show all data. [7.1.2]

14. Record data for 5 minutes (or until the oxygen level stabilizes). Then, stop data recording. [6.2]

15. Remove the oxygen-sensing element from the sampling bottle.

16. Dispose of the contents of the bottle as directed and rinse the bottle thoroughly.

17. Repeat Steps 9 through 16 with the chilled yeast suspension.

18. Repeat Steps 9 through 16 with the heated yeast suspension.

19. Save your experiment, [11.1] and clean up the equipment and work area.

Organize Data

1. **Organize Data** Using available tools on your data-collection system find values for the initial and final oxygen gas percentages.[9.1] Calculate the change in oxygen (%); then calculate the rate of change, where the Rate of change = [Change in oxygen (%)] ÷ 5 minutes. Record these in the Data Table.

Oxygen Measurements at Room Temperature, and for Chilled and Heated Yeast				
	Initial O_2 Gas (%)	Final O_2 Gas (%)	Change in O_2 (%)	Rate (per minute)
Room Temp.				
Chilled				
Heated				

2. **Graph** On graph paper, graph your data for Oxygen Concentration versus Time. Label the overall graph, the x-axis, the y-axis, and include units on the axes. Label the data for each trial: room-temperature, chilled, and heated yeast.

Analyze and Conclude

1. **Infer** Why were oxygen levels measured?

_____

_____

_____

2. **Interpret Graphs** Which sample of catalase produced the greatest oxygen concentration? What does that indicate about the enzyme activity?

_____

_____

_____

3. **Draw Conclusions** Describe the effect that heating the catalase had on the reaction with the hydrogen peroxide. How did the effect of cooling compare to the effect of boiling the catalase? How can you explain the difference between these two trials?

_____

_____

_____

4. **Apply Concepts** Why are enzymes important in biological systems? What would happen to cells if enzymes were not present?

_____

_____

_____

5. **Infer** Using what you have observed about enzymes at high temperatures, explain why fevers over 40.5°C (105°F) are so dangerous in humans.

_____

_____

_____

Extend Your Inquiry

How can pH affect catalase activity? Design an experiment to test your hypothesis. Ask your teacher for comments on your design. If necessary, revise the procedure. With your teacher's permission and supervision, carry out your experiment.

Lab 2 **Microclimates**

Problem

What factors affect a microclimate?

Introduction

Take a walk around Savannah, Georgia, and within a few blocks the temperature may go from 23°C to 27°C. Drive 30 miles inland and the temperature may change from the original 23°C to 34°C. What makes it possible for the temperature to vary by more than 10°C over a relatively short distance? Buildings, hills, plants, and water are all responsible for creating microclimates in and around Savannah. A microclimate occurs when environmental conditions within a small distance vary significantly from the climate of the surrounding area. Many microclimates are so small that humans may not even notice them, but they are still important to other organisms. For example, beneath a fallen tree, insects may find a patch of a cool, moist soil, ideal for hiding from predators.

In this lab, you will measure the temperature at different locations around your school. You will make observations about the different locations and draw conclusions about what factors affect microclimates near your school.

Skills Focus

Observe, Draw Conclusions, Predict, Use a Data-Collection System and Sensor

Materials

- mobile PASCO data-collection system
- temperature sensor

Safety ⚠

Be aware of dangerous plants that are common in your area. Do not touch or approach animals, because they may bite. Do not go into restricted areas or near moving vehicles.

Pre-Lab Questions

1. **Relate Cause and Effect** Why is it important to consistently hold the sensor 1 meter above the ground?

   _____

   _____

   _____

2. **Predict** Which of your locations do you think will have the coolest temperature? Which will have the warmest?

   _____

   _____

   _____

3. **Relate Cause and Effect** What physical structures in or around your schoolyard might influence the data?

   _____

   _____

   _____

Procedure

Part A: Setup

1. Your teacher will provide you with the current weather conditions in your area, or you can obtain them from the radio, newspaper, TV, or Internet. These values are your control data. Record your control data in the Data Table in the Organize Data section of this lab.

2. Start a new experiment on the data-collection system.[1,2]

3. Connect a temperature sensor to your data-collection system.

4. Create a data display in a graph showing Temperature on the y-axis and Time on the x-axis.[7.1.1]

5. Select four locations near your school to take temperature measurements.

   Location 1: _____

   Location 2: _____

   Location 3: _____

   Location 4: _____

6. Show your list of locations to your teacher for approval.

Part B: Collect Data

7. After obtaining your teacher's permission, go to your first location to collect data.

8. Write a description of the physical conditions of this location (shade, sun, asphalt, moist, etc.) in the Data Table.

9. Hold your sensor approximately 1 meter above the ground, and start data recording. Collect data for 60 seconds. You can monitor the time using the graph screen. Stop recording data after 60 seconds.[6.2]
   NOTE: Cover the sensor to shield it from direct sun.

10. Name the data run[8.2] with a location description, such as "Football Field," and write the name in the Data Table.

11. Repeat Steps 7–10 at the other three locations.

Organize Data

1. **Organize Data** View your statistics of data to complete the Data Table.[9.4]

Data Table		
Location	Description of Location (shade, sun, asphalt, concrete)	Average Temperature (°C)
Local weather conditions (control data)		

2. **Graph** On graph paper, plot a bar graph that represents the average temperature recorded at each location. Be sure to give your graph a title and to label the axes correctly (remember to include units).

Analyze and Conclude

1. **Interpret Data** At which location was the temperature the highest? The lowest? What could account for the differences between these locations?

_____

_____

_____

_____

2. **Infer** Compare your predictions to your results. Were your predictions accurate, or did your results show something other than what you expected? Explain any differences.

_____

_____

_____

_____

3. **Draw Conclusions** If your temperature data showed differences between locations, what factors in the natural environment may have influenced these changes? What human activities may have influenced them?

_____

_____

_____

_____

4. **Relate Cause and Effect** What effect do plants have on the microclimate?

_____

_____

_____

Extend Your Inquiry

Repeat the activity at different points throughout the school year. Are the changes in microclimates greater in warm or cold conditions? You could also test differences in temperature by taking measurements at different distances from the ground. How are these differences similar to changes in elevations?

Lab 3 **Respiration in Seeds**

Problem

What factors affect the rate of respiration in seeds?

Introduction

When a seed becomes moist, the seed coat absorbs water, softens, and allows water to penetrate the seed. As soon as the cells inside the seed absorb the water, enzymes are activated, and the seedling begins the process of cellular respiration. Cellular respiration converts glucose ($C_6H_{12}O_6$), oxygen (O_2), adenosine diphosphate (ADP), and inorganic phosphate into carbon dioxide (CO_2), water (H_2O), and adenosine triphosphate (ATP). The process provides the new plant with ATP molecules, allowing it to grow.

In this lab, you will measure the rate of cellular respiration occurring in seeds. To do this, you will measure the amount of carbon dioxide gas—one of the products of respiration—produced by seeds over time. You will compare the respiration rate of dry seeds to that of germinating seeds, and you will determine whether temperature affects the rate of respiration.

Skills Focus

Observe, Predict, Use a Data-Collection System and Sensor

Materials

- PASCO data-collection system
- carbon dioxide gas sensor
- sensor extension cable
- sampling bottle (included with sensor)
- beaker, 1-L
- 10 dry pea or bean seeds
- 10 pea or bean seeds soaked in room-temperature water
- 10 pea or bean seeds soaked in ice water
- ice, cubed or crushed
- water

Safety

Always wear goggles. Wash your hands thoroughly with soap and warm water at the end of the lab.

Pre-Lab Questions

1. **Recall** What is the chemical equation for cellular respiration?

   _____

2. **Relate Cause and Effect** In Step 7, why is it important to create a seal by firmly inserting the sensor into the bottle?

   _____

   _____

   _____

3. **Predict** Which seeds will likely have the highest rate of respiration? Explain your answer.

   _____

   _____

   _____

   _____

   _____

Procedure

Part A: Setup

1. Place 400 mL of cubed or crushed ice into a 1-L beaker. Add 400 mL of water to the beaker. Add 10 germinating seeds into the ice water. Allow the seeds to soak for at least 20 minutes. Move on to Step 2 while seeds soak.

2. Start a new experiment on the data-collection system.[1.2]

3. Connect the extension cable into a port on the data-collection system. Connect the other end of the extension cable to the carbon dioxide sensor.[2.1]

4. Display the carbon dioxide Concentration in parts per million (ppm) on the y-axis and Time in seconds (s) on the x-axis of a graph.[7.1.1]

5. Calibrate the carbon dioxide sensor.[3.1]

6. Place 10 dry, dormant seeds into the sampling bottle.

7. Create a seal by putting the end of the carbon dioxide gas sensor into the sampling bottle. Firmly plug the end of the sampling bottle with the rubber stopper.

Figure 1 Complet

Part B: Collect Data

8. Record data for 10 minutes. Then, stop data recording.[6.2]

9. Name the data run Dormant Seeds.[8.2]

10. Save your experiment.[11.1]

11. Remove the carbon dioxide gas sensor.

12. Remove the seeds from the sampling bottle.

13. Fill the sampling bottle with water. Then empty the water from the bottle.

14. Repeat Steps 2–13 with the seeds soaked in water at room temperature. In Step 9, name the data run Germinating Seeds.

15. Repeat Steps 2–13 with the seeds soaked in ice water. In Step 9, name the run Cold, Germinating Seeds.

16. Save your experiment[11.1] and clean up the equipment and work area. Dispose of seeds as directed by your teacher.

Organize Data

1. **Organize Data** Using available tools on your data collection system, find the values for initial and final carbon dioxide concentrations (ppm).[9.1] Calculate the rate of change for each run of data collected by dividing the carbon dioxide concentration (in ppm) by the time (in seconds). Record these values in the Data Table.

Data Table				
Run	Initial CO_2 Concentration (ppm)	Final CO_2 Concentration (ppm)	Time (s)	Rate of CO_2 Production (ppm/s)
Dry, dormant seeds			600	
Room-temperature, germinating seeds			600	
Cold, germinating seeds			600	

2. **Graph** On graph paper, graph your data for carbon dioxide concentration over time. Give your graph a title, and be sure to label the scale of the *x*-axis and the *y*-axis. Label the data for Dry, Dormant Seeds; Room-Temperature, Germinating Seeds; and Cold, Germinating Seeds.

Analyze and Conclude

1. **Compare and Contrast** How do the rates of carbon dioxide production for room-temperature, germinating seeds; dry, dormant seeds; and cold, germinating seeds compare?

_____

_____

_____

2. **Predict** If you add water to the dry seeds, what will happen to the concentration of carbon dioxide inside the flask?

_____

_____

_____

3. **Explain** Why do plants need to perform cellular respiration even though they are photosynthetic organisms?

_____

_____

_____

4. **Draw Conclusions** What are some factors that affect the rate of cellular respiration in an organism? Include evidence you collected in this lab activity.

_____

_____

_____

Extend Your Inquiry

Repeat the activity with different kinds of seeds, such as barley or sunflower seeds. Or test seeds under different conditions. Would seeds that germinated 24 hours prior to the experiment have a different rate of respiration compared seeds that germinated 72 hours prior to the experiment? What effect might you observe if the water used to moisten the seeds had a different pH?

Lab 4 Transpiration

Problem

Does transpiration take place faster on a windy day or a day when the air is still?

Introduction

Water transport is an essential process in land plants. Water is obtained by roots at the bottom of the plant but is used by leaves at the top. Efficient transport of water between the roots and leaves is needed for the plant's survival.

Transpiration is the loss of water through leaves. As each water molecule evaporates from the underside of a leaf, cohesion pulls more water molecules out from the xylem vessels in the leaf. This pull is transmitted all the way down from the leaves to the root in a continuous water column, allowing the efficient movement of water from the roots up to the leaves. The amount of water needed by plants for growth and maintenance of tissue is small compared to the amount that is lost through the transpiration. But, if water lost from leaves is not replaced by water transported up from the plant roots, the plant will wilt and die.

Any factor that affects a plant's transpiration rate affects water transport throughout the plant. This lab will measure transpiration rates in still air and moving air. The data will be collected by measuring pressure changes as the plant takes up water into the stem.

Skills Focus

Observe, Predict, Use a Data-Collection System and Sensor

Materials

- PASCO data-collection system
- barometer/low-pressure sensor, including plastic tube
- sensor extension cable
- large base and support rod
- clamp, utility
- clamp, three-finger
- pipette

- knife
- fan
- bowl
- glycerin, 1 mL
- petroleum jelly, 2 to 3 g
- plant seedling, 12 to 15 cm tall
- water, 1 L

Safety

Always wear goggles. Do not direct the knife toward yourself or others. Use the knife only as instructed. Do not allow the barometer/low-pressure sensor to get wet. Wash your hands thoroughly with soap and warm water at the end of the lab.

Pre-Lab Questions

1. **Apply Concepts** Why do you think it is important to keep the plant stem under water after you cut it?

_____

_____

_____

_____

2. **Relate Cause and Effect** Why is petroleum jelly needed? What would happen if the petroleum jelly came in contact with the cut end of the stem?

_____

_____

_____

_____

3. **Apply Concepts** Do you expect to see an increase or decrease in pressure as transpiration occurs? Explain your answer.

_____

_____

_____

_____

_____

4. **Predict** Do you expect the rate of transpiration to be greater in still air or moving air? Explain your answer.

_____

_____

_____

_____

_____

Procedure

Part A: Setup

1. Start a new experiment on the data-collection system.[1,2]

2. Using a sensor extension cable, connect a barometer/low-pressure sensor to the data-collection system.[2.1]

3. Display a graph on the data-collection system showing Barometric Pressure (hPa) on the y-axis and Time (s) on the x-axis.[7.1.1]

4. Put a drop of glycerin on the barbed end of the quick-release connector and insert the barb into one end of the plastic tube. The barb will not go completely into the tube. Be sure that it is in far enough so it does not easily fall out.

5. Use a knife to cut the stem of a plant seedling 2 to 3 cm above the soil. Immediately immerse the cut end of the seedling in a bowl of water. With the plant submerged, carefully cut a 45° angle near the base of the stem.

6. Bend the piece of plastic tube into a U shape. Use a pipette or an eyedropper to fill the tube with water.

7. Place the tube into the bowl of water. While holding the tube underneath the water, insert the plant stem, cut-end first, into the tube. **NOTE:** Avoid creating air bubbles in the tube. If you notice any air bubbles in the tube, start over.

8. Raise the tube and plant out of the water, keeping the tube in a U shape. Adjust the level of the tube so there is a 2- or 3-cm air gap beneath the connector end. Water will spill out of the tube from the end opposite the plant. Be sure that water in the tube *always* stays in contact with the plant. **NOTE:** Don't allow the connector-end of the tube to be lower than the plant-end of the tube.

9. Carefully dab petroleum jelly around the end of the tube to create an airtight seal around the edge of the plant stem. If water is still leaking from the tube, add more petroleum jelly. **NOTE:** Be sure that the petroleum jelly does not come in contact with the cut end of the plant stem. If air bubbles form around the cut end of the stem, pull the tube away from the stem. Use the eyedropper to refill the open end of the tube with water. Put the stem back into the water in the tube.

10. Secure the tube with the plant seedling in an upright position using a clamp and the base-and-support rod.

Figure 1 Tube setup

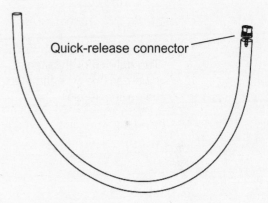

Quick-release connector

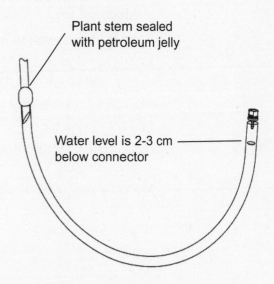

Plant stem sealed with petroleum jelly

Water level is 2-3 cm below connector

11. Mount the barometer/low-pressure sensor to the support rod with the three-finger clamp. The pressure port should be above the cut end of the stem, preventing water from entering the sensor, as shown in Figure 2. **NOTE:** Make sure that no water enters the sensor. There should be a 2- to 3-cm air pocket between the water level and the pressure port.

Figure 2 Completed setup

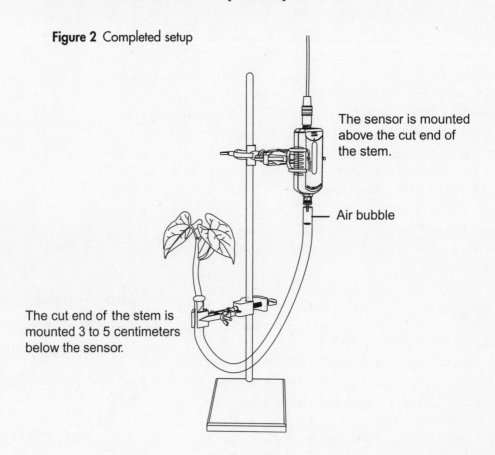

The sensor is mounted above the cut end of the stem.

Air bubble

The cut end of the stem is mounted 3 to 5 centimeters below the sensor.

12. Align the quick-release connector on the tube with the connector on the pressure port of the sensor. Push the connector onto the port, and then turn the connector clockwise until it clicks (about one-eighth turn).

Part B: Transpiration in Still Air, Collect Data

13. Start recording data.[6.2] **NOTE:** Do not move the sensor up or down on the support rod while recording data.

14. Adjust the scale of the graph to show all data.[7.1.2]

15. Record data for 600 seconds.[6.2]

16. To restore the pressure in the tube to the original value, carefully disconnect and then reconnect the tube to the sensor.

17. Name run 1 Still Air.[8.2]

Part C: Transpiration in Moving Air, Collect Data

18. Place a fan about 1 meter from the plant seedling.

19. Put the fan on a low setting so it blows a light breeze over the seedling.

20. Repeat Steps 13–15.

21. Name run 2 Wind.[8.2]

22. Save your experiment,[11.1] and clean up according to your teacher's instructions.

Organize Data

1. **Organize Data** Use the graph tools of your data-collection system [9.2] to find the data from your graph to complete the first two columns in the Data Table. Then, calculate the rate of transpiration.

Data Table			
Transpiration Run	ΔP (hPa)	Δt (s)	Rate of Transpiration (hPa/min)
Without fan			
With fan			

2. **Graph** On graph paper, plot each of your runs—with fan and without fan. Give your graph a title and be sure to label the scale of the x-axis and the y-axis.

Analyze and Conclude

1. **Analyze Data** What was the rate of pressure change in the plastic tube when the fan was on? When it was off? Does a decrease in pressure in the tube correspond to an increase or to a decrease in water loss through the seedling's stomata? Explain.

_____

_____

_____

_____

_____

2. **Analyze Data** Did the fan affect the rate of pressure change? Explain how the fan affects transpiration. What natural condition does the fan mimic?

_____

_____

_____

3. **Applying Concepts** Will all plant species have the same transpiration rate under similar environmental conditions? Why or why not?

_____

_____

_____

4. **Predict** How would covering the lower side of the leaves with petroleum jelly affect transpiration?

_____

_____

_____

5. **Applying Concepts** Would the transpiration rate in a conifer leaf (needle) be higher or lower than in a deciduous (broad) leaf? Explain your answer.

_____

_____

_____

Extend Your Inquiry

Design an experiment to test how other environmental factors such as light intensity, humidity, and temperature can affect the rate of transpiration.

Lab 5 **Regulation of Body Heat**

Problem

How do external conditions affect your body's internal conditions?

Introduction

On a cold day, you start to shiver. On a hot day, you start to sweat. Why does your body react this way? Both shivering and sweating are mechanisms that maintain your body's core temperature. To do this, your nervous system must be able to detect changes and then adjust body processes to maintain relatively stable and constant internal conditions. This state is called homeostasis.

In this lab, you will test the effects of external conditions on your body's temperature. To do this, you will use a temperature sensor to measure the surface temperature of your hand under various conditions.

Skills Focus

Observe, Predict, Use a Data-Collection System and Sensor

Materials

- PASCO data-collection system
- quad temperature sensor
- large bowl (or similar container)
- fan
- tape or adhesive covers
- ice, cubed or crushed, 1 L
- water, 1 L
- towel (several)
- glove or mitten

Safety 🧤 🥽

If the ice water causes discomfort or pain, stop data collection. Do not allow any electric cords to touch the water.

Pre-Lab Questions

1. **Relate Cause and Effect** Why is it necessary to sit still during Part B of this procedure?

_____

_____

2. **Controlling Variables** Why is the temperature recorded for still air called the "control"?

_____

_____

3. **Predict** Which will probably change skin temperature more: a hand with room-temperature air moving across it or a hand held in ice water?

_____

_____

Procedure

Part A: Setup

1. Start a new experiment on the data-collection system.[1.2]

2. Connect the two temperature sensors into two ports on the data-collection system.[2.1 & 2.2]

3. Display Temperature in degrees Celsius (°C) on the y-axis and Time in seconds (s) on the x-axis of a graph.[7.1.1]

4. Attach a temperature sensor to the back of each hand. Fasten the sensors using tape or adhesive covers, as shown in Figure 1.

Figure 1 Temperature Sensor on Hand

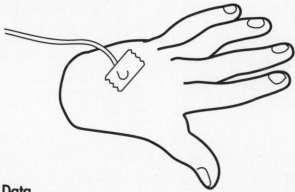

Part B: Collect Data

5. During data collection, the person who is being measured should be sitting and relaxed. The person should not look at the data as it is being recorded. Start data recording.[6.2] Record data for 6 minutes.

6. While data is collecting, adjust the scale of the graph to show all data.[7.1.2]

7. Name data run 1 Still Air.[8.2]

8. Place a fan so that it will blow air across the right hand and *not* blow air across the left hand. Start the fan.

9. Repeat Steps 5 and 6.

10. Name data run 2 Moving air.[8.2]

11. Wait 5 minutes to allow your hand to return to a normal temperature before doing Step 12.

12. Place a glove or mitten on the right hand. Start data recording for 6 minutes.

13. After 2 minutes, remove your hand from the glove. Continue to record data for 4 more minutes.

14. Repeat Step 6.

15. Name data run 3 Glove or Mitten.

16. Wait 5 minutes to allow your hand to return to a normal temperature before proceeding.

17. Put ice and water into a large bowl. Leave enough room in the container so that water will not spill when you place your hand into the ice water in Step 18.

18. Place your right hand in the ice water so that your palm and fingers are completely in the water. **IMPORTANT:** Make sure to leave the back of your hand and the sensor above water, as shown in Figure 2.

Figure 2 Placement of hand in ice water

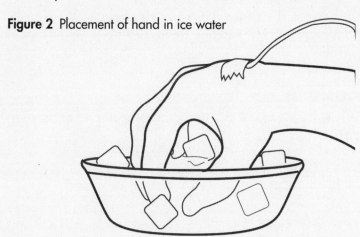

19. After 2 minutes, remove your hand from the ice water. Continue collecting data for 4 more minutes.
    CAUTION: If at any point before 2 minutes has passed, it becomes too uncomfortable, remove your hand from the ice water.

20. Repeat Step 6.

21. Name data run 4 Ice Water.[8.2]

22. Save your experiment[11.1] and clean up according to your teacher's instructions.

Organize Data

1. **Organize Data** Use available tools on your data-collection system [9.1 & 9.2] to complete the following table.

	Left Hand (control, always in still air)			Right Hand		
Trial	Starting Temperature (°C)	Temperature at Time of Greatest Change (°C)	Greatest Change in Temperature (°C)	Starting Temperature (°C)	Temperature at Time of Greatest Change (°C)	Greatest Change in Temperature (°C)
Still Air						
Moving Air						
Glove						
Ice Water						

Data Table

2. **Graph** On graph paper, plot each of your runs—control, fan, glove, and ice water. Give each of your graphs a title, and be sure to label the scale of the *x*-axis and the *y*-axis.

Analyze and Conclude

1. **Draw Conclusions** What would cause a change in the temperature of skin exposed to moving air?

_____

_____

_____

_____

2. **Analyze Data** What happened to the skin temperature during the first minute after the hand was removed from the ice water? What caused this change?

_____

_____

_____

_____

_____

3. **Evaluate Design** How can you explain any differences in the starting temperatures of each hand? How could you reduce this problem?

_____

_____

_____

_____

_____

4. **Relate Cause and Effect** How do data show your body is working to maintain homeostasis?

_____

_____

_____

_____

5. **Apply Concepts** What processes might occur in the body that allow a person to maintain a relatively constant core body temperature, even if the person gets frostbite on his or her fingers and toes?

_____

_____

_____

_____

Extend Your Inquiry

Design an experiment to test how other external factors can affect skin temperature. Or, test which of these affects skin temperature more—water that is very cold but still or water that is only moderately cold but is moving over the skin.

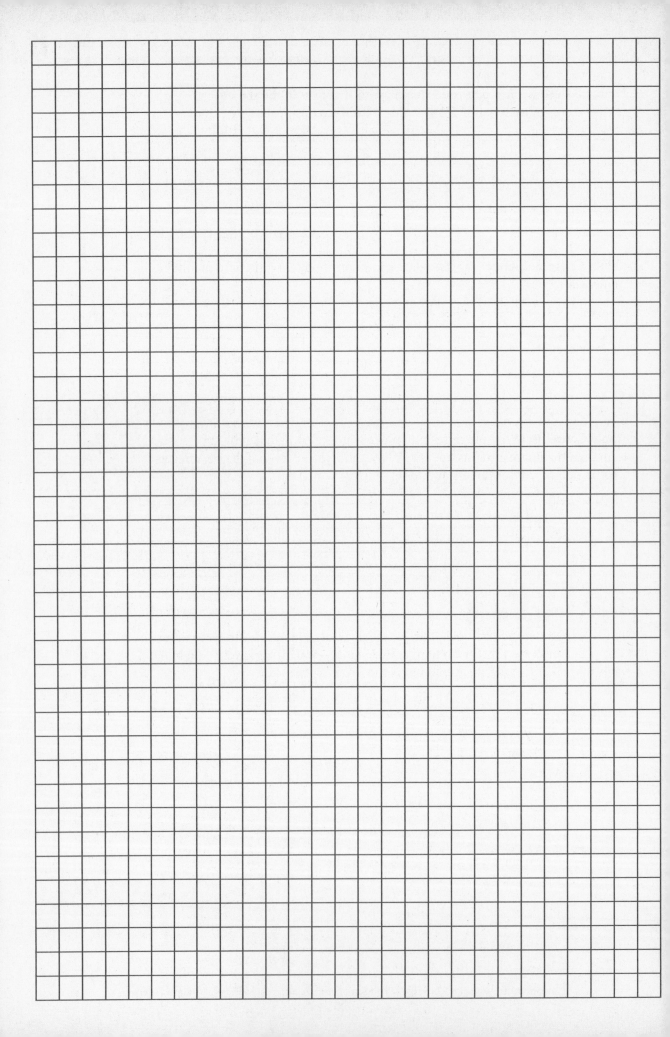

Vernier Labs

Teacher Notes and Answers are on the CD-ROM.

The lab activities from Vernier are listed in the table below along with the sensor required to conduct each lab.

Lab Activity	Probeware
Enzyme Action	Vernier O_2 Gas Sensor
Testing Watersheds	Vernier Dissovled O_2 Sensor and calibration bottle , Vernier pH Sensor, Vernier Conductivity Sensor, Vernier Temperature Sensor
Comparing Respiration Rates in Seeds	Vernier CO_2 Gas Sensor
Transpiration Rates	Vernier Gas Pressure Sensor
Human Respiration	Vernier Respiration Belt

The CD-ROM provided in the front of the book contains both a PDF and an editable copy of the student version of each lab. In addition, the CD-ROM includes a teacher's version with an answer guide and teaching tips. Finally, the CD-ROM includes information about the use of individual data-collection systems and sensors specific to each lab. Please refer to the Vernier Technical Guide found in he Vernier folder on the CD-ROM.

These labs have been adapted from Experiments 6A Enzyme Action, 20 Watershed Testing, 11B Cell Respiration, 10 Transpiration, and 26 Control of Human Respiration from the *Biology with Vernier Lab Manual* by Kelly Redding and David Masterman.

Lab 6 # Enzyme Action

Problem

How does concentration and temperature affect the rate of an enzyme-catalyzed reaction?

Introduction

Enzymes act as catalysts, increasing chemical reaction time in living organisms. For an enzyme to function properly, temperature is very important. Most organisms have a specific temperature range in which they survive, and their enzymes function best within that temperature range. Outside of that range, enzymes may cease to function.

Hydrogen peroxide (H_2O_2) is toxic to most living organisms. Many organisms decompose hydrogen peroxide with enzymes before it can do much damage, as it can be broken down into two harmless products—water (H_2O) and oxygen (O_2). Although this reaction occurs spontaneously, enzymes increase the rate of reaction. The enzyme catalase helps increase the rate at which hydrogen peroxide is broken down.

In this lab, you will measure the rate of enzyme activity under two conditions: different enzyme concentrations and at different temperatures.

Skills Focus

Use Sensors and a Data-Collection Interface, Analyze Data, Infer

Materials

- Vernier data-collection interface[1]
- Vernier O_2 Gas Sensor
- 400-mL beaker
- 10-mL graduated cylinder
- 18 × 150 mm test tubes (3)
- 250-mL Nalgene bottle
- 3.0% hydrogen peroxide solution
- yeast suspension
- ice
- test tube rack
- thermometer
- 3 dropper pipettes
- hot plate
- Logger *Pro* (optional)

Safety 🜸 🜲 🜳 🜴 🜵 🜶 🜷

Always wear goggles and an apron when using a water bath. If you have glass beakers or test tubes, check for cracks or chips. Alert your teacher if you break a glass object. Avoid contact with the surface of the hot plate. To avoid electric shocks, make sure cords, connections, and your hands are dry. Wash your hands thoroughly with soap and warm water at the end of the lab.

Pre-Lab Questions

1. **Relate Cause and Effect** How will you know that a chemical reaction is taking place in Part B and Part C?

   _____

   _____

2. **Predict** Which concentration of yeast suspension do you think will increase the reaction time the most? Why?

   _____

   _____

   _____

3. **Relate Cause and Effect** Why are you measuring oxygen levels?

   _____

   _____

   _____

Procedure

Part A: Set Up the Experiment

1. Obtain and wear goggles.

2. Prepare the data-collection interface. Set the data-collection rate to 0.2 samples/second and the data-collection length to 180 seconds.[8]

3. Label three test tubes 1, 2, and 3. Fill each test tube with 5 mL of the 3.0% hydrogen peroxide solution and 5 mL of water. Place each test tube on the test tube rack.

Part B: Testing the Effect of Enzyme Concentration

4. Using a clean dropper pipette, add 5 drops of yeast suspension to test tube 1. Cover the opening of the test tube with a gloved finger and gently invert the test tube two times. Pour the contents of the test tube into a clean 250-mL Nalgene bottle. Place the O_2 Gas Sensor into the bottle as shown in Figure 1. Gently push the sensor down into the bottle until it stops. The sensor is designed to seal the bottle with minimal force. After 30 seconds, start data collection.[2]

Figure 1 Completed Setup

5. After 3 minutes, remove the O_2 Gas Sensor from the Nalgene bottle. Rinse the bottle with water and dry it with a paper towel.

6. Save your data.[3]

7. Repeat Steps 3–6, adding 10 drops of yeast suspension to test tube 2.

8. Repeat Steps 3–6, adding 20 drops of yeast suspension to test tube 3.

9. Find the value of the slope, m, for the change in oxygen concentration for each of the three reactions.[4] The slope is the rate of the reaction. In Data Table 1, record the value of the slope, m, for each of the three solutions.

Data Table 1		
Test Tube	Yeast Concentration	Slope, or Rate of Reaction (%/s)
1	5 drops	
2	10 drops	
3	20 drops	

Part C: Testing the Effect of Temperature

As a class you will test the effects of temperature on the enzyme. Each group will test a specific temperature range and share its data with the rest of the class. Using your assigned temperature range, set up your water bath as described below. Place a thermometer in your water bath to assist in maintaining the proper temperature.

- 0–5°C: Fill a 400-mL beaker with ice and water.
- 20–25°C (room temperature): No water bath is needed.
- 30–35°C: Fill a 400-mL beaker with very warm water; heat it on a hot plate.
- 50–55°C: Fill a 400-mL beaker with hot water; heat it on a hot plate.

10. Rinse the three numbered test tubes used for Part I. Fill each test tube with 5 mL of 3.0% H_2O_2 and 5 mL of water, and then place the test tubes in the water bath. The test tubes should be in the water bath for 5 minutes before proceeding to Step 12. Record the temperature of the water bath, as indicated on the thermometer, in the space provided in Data Table 2.

11. Repeat Step 2.[8]

12. Add 10 drops of the enzyme solution to each test tube. For each test tube, repeat Steps 4–6.

13. Record the reaction rates in Data Table 2. Calculate the average reaction rate for the three trials you tested. Record the average rate in Data Table 2.

Data Table 2	
Test Tube Label	Slope, or Rate (%/s)
Trial 1	
Trial 2	
Trial 3	
Average	
Temperature range: °C– °C	

14. Record the average rate and the temperature of your water bath from Data Table 2 on the class chalkboard. When the entire class has reported its data on the chalkboard, record the class data in Data Table 3.

Data Table 3 (Class Data)		
Temperature Range	Actual Temperature	Average Rate of Reaction (%/s)
0–5°C range		
20–25°C range		
30–35°C range		
50–55°C range		

15. **Graph** Using the data from Data Table 3, make a graph of the rate of enzyme activity versus temperature by hand or using Logger *Pro* software.[5] Plot the rate values from the class data on the *y*-axis and the temperature on the *x*-axis.

16. Follow your teacher's instructions for dismantling the setup and for cleanup. Wash your hands thoroughly with soap and warm water before leaving the lab.

Analyze and Conclude

1. **Interpret Tables** Refer to Data Table 1. Which concentration of yeast had the highest reaction rate? Which concentration of yeast had the lowest rate?

_____

_____

_____

2. **Predict** If you increased the concentration of the enzyme solution to 30 drops, what do you think will happen to the rate of reaction?

_____

_____

_____

3. **Interpret Graphs** Refer to your graph. At what temperature is the rate of enzyme activity the highest? Lowest? What does this indicate about enzyme activity?

_____

_____

_____

4. **Apply Concepts** Why is it important for your body to maintain a stable internal temperature?

_____

_____

_____

Extend Your Inquiry

How can pH affect catalase activity? Design an experiment to test your hypothesis. Ask your teacher for comments on your design. If necessary, revise the procedure. With your teacher's permission and supervision, carry out your experiment.

Lab 7 **Testing Watersheds**

Problem

How do you determine water quality?

Introduction

On a hot summer day at Arizona's Slip Rock State Park, a sign states "Unsafe water, no swimming." Every day at the park the water quality is tested to ensure that bacteria levels are not at a harmful level. In addition to knowing if it is safe to swim, there are many reasons for determining water quality. We depend on fresh water for drinking and growing crops. Pollutants in the water can affect an entire watershed.

How is water quality measured? In the 1970s, the National Sanitation Foundation developed a standard index for measuring water quality. The Water Quality Index, or WQI, is a series of nine tests that measures temperature, pH, turbidity, total solids, dissolved oxygen (D.O.), biochemical oxygen demand, phosphates, nitrates, and fecal coliform. Through these nine tests, the water quality value, or Q-value, of the water is determined.

In this lab, you will conduct a modified version of the WQI tests. You will test only water temperature, dissolved oxygen, pH, and total dissolved solids. This modified version of the WQI will allow you to determine the general quality of the stream or lake you are sampling.

Materials

- Vernier data-collection interface[1]
- Vernier Dissolved Oxygen Sensor
- Vernier pH Sensor
- Vernier Conductivity Sensor
- Vernier Temperature Sensor
- 4 water sampling bottles
- large plastic cup or beaker
- D.O. calibration bottle
- D.O. filling solution
- distilled water
- permanent marker

Skills Focus

Measure, Analyze Data, Use Sensors and a Data-Collection Interface

Safety 🔥 🦺 🧪

Be aware of dangerous plants that are common in your area. Be careful in and near the water. Do not enter any fast-moving streams. After performing the lab, wash your hands with warm water.

Pre-Lab Questions

1. **Apply Concepts** In Step 6, why is it important to keep the bottle under water when you add the stopper?

   _____

   _____

   _____

2. **Infer** Think about the area where you will be testing water quality. Are there any sources of pollution that you know of? How might pollution affect your results?

   _____

   _____

   _____

Procedure

Part A: Set Up Dissolved Oxygen Sensor, Conductivity Sensor, and Data-Collection Interface

1. To prepare the Dissolved Oxygen Sensor, remove the protective cap. Unscrew the membrane cap from the tip of the sensor. Using a pipet, fill the membrane cap with 1 mL of D.O. Electrode Filling Solution. Carefully thread the membrane cap back onto the electrode. Place the sensor into a 250 mL beaker containing distilled water. Connect the sensor to your data-collection interface.

2. Set the switch on the Conductivity Sensor box to 0–2000 µS/cm. Connect the Conductivity Sensor, Temperature Sensor, and pH Sensor to your data-collection interface. If you have older sensors that do not auto-ID, manually set up the sensors.

3. Prior to taking any readings, allow the Dissolved Oxygen Sensor to warm up. With the sensor still in the distilled water, wait 10 minutes. The sensor must stay connected at all times to function.

4. Calibrate the Dissolved Oxygen Sensor according to your teacher's instruction.[9]

Part B: Collecting Samples

You will take four water samples to measure the dissolved oxygen concentration, pH, water temperature, and total dissolved solid (TDS) concentration at two locations. Your teacher will select two sites at which you will make measurements.

5. Carefully enter the water and move away from the edge as long as it remains safe. Be careful of slippery or sharp rocks and fast-moving water. **CAUTION:** If the current is strong, do not enter the water.

6. Rinse the sampling bottle several times with stream water. Place the bottle below the surface and allow water to flow into the opening for three minutes. Fill the bottle so it is completely full and place the stopper in the bottle while the bottle is still under water. Label the sample bottle Site 1a.

7. Take your sample away from the water to make measurements.

8. Remove the Dissolved Oxygen Sensor from its storage bottle. Place the sensor into the water and gently swirl to allow water to move past the sensor's tip. When the reading is stable, record the dissolved oxygen concentration in Data Table 1. Remove the sensor from the water and place it back into the storage bottle. **IMPORTANT:** Place only one sensor in the water at a time.

9. Place the Temperature Sensor into the water sample. When the reading has stabilized, record the temperature in Data Table 1. Remove the sensor from the water.

10. Remove the pH Sensor from its storage bottle. Place the sensor into the water sample. When the reading has stabilized, record the pH in Data Table 1. Remove the sensor from the water and place it back into the storage bottle.

11. Place the tip of the Conductivity Sensor into the water sample. The hole near the tip of the sensor should be submerged completely. When the reading has stabilized, record the TDS value in Data Table 1. Remove the sensor from the water.

12. Choose a spot 6 meters from Site 1a. Repeat Steps 6–11. Label it as Site 1b.

13. Repeat Steps 5–12 at a location away from from Site 1. Label them as Site 2a and Site 2b.

Part C: Collect Data and Determine Water Quality Index

14. Calculate the averages for measurements at Sites 1 and 2, and record the results in Data Table 1.

Data Table 1				
Location	Temperature (°C)	Dissolved Oxygen (mg/L)	pH	Total Dissolved Solids (mg/L)
Site 1a				
Site 1b				
Average				
Site 2a				
Site 2b				
Average				
Temperature Difference				

15. Record the average levels of dissolved oxygen that you measured at each site in Data Table 2.

16. Determine the barometric pressure in mm Hg by using either a barometer or Table 9: Approximate Barometric Pressure at Different Elevations.

17. Use barometric pressure and temperature values to determine the dissolved oxygen saturation. Look up the level of dissolved oxygen for air saturated water in mg/L from Table 10: 100% Dissolved Oxygen Capacity. Record the value for each site in Data Table 2.

Data Table 2 Dissolved Oxygen (% saturated)			
	Average Dissolved Oxygen (mg/L)	Dissolved Oxygen in Saturated Water	% Saturated
Site 1			
Site 2			

18. Determine the % saturation by using the following formula:

$$\% \text{ saturation} = \frac{\text{Measured D.O. level}}{\text{Saturated D.O. level}} \times 100$$

Record the % saturation of dissolved oxygen in Data Table 2.

19. Using Tables 3–5, determine the Q-value for dissolved oxygen, pH, and TDS. Use the number closest to your value to obtain the closest Q-value. Record your result in Data Table 7 for Site 1 and in Data Table 8 for Site 2.

Table 3	
Dissolved Oxygen (D.O.) Test Results	
D.O. (% saturation)	Q-value
0	0
10	5
20	12
30	20
40	30
50	45
60	57
70	75
80	85
90	95
100	100
110	95
120	90
130	85
140	80

Table 4	
pH Test Results	
pH	Q-value
2.0	0
2.5	1
3.0	3
3.5	5
4.0	8
4.5	15
5.0	25
5.5	40
6.0	54
6.5	75
7.0	88
7.5	95
8.0	85
8.5	65
9.0	48
9.5	30
10.0	20
10.5	12
11.0	8
11.5	4
12.0	2

20. Find the difference between the average temperatures of Sites 1 and 2. Record the difference in the Data Table 1.

21. Use Table 6 and the temperature difference to determine the Q-value for temperature. Record your result in both Data Table 7 and Data Table 8.

22. Multiply each Q-value by the given weighting factor. Record the total Q-value in Data Tables 7–8.

Table 5	
Total Dissolved Solids (TDS) Test Results	
TDS (mg/L)	Q-value
0	80
50	90
100	85
150	78
200	72
250	65
300	60
350	52
400	46
450	40
500	30

Table 6	
Temperature Test Results	
Temperature Difference (°C)	Q-value
0	95
5	75
10	45
15	30
20	20
25	15
30	10

23. Determine the overall water quality of your stream by adding the four total Q-values in Data Table 7 for Site 1 and in Data Table 8 for Site 2. Record the result. The closer this value is to 100, the better the water quality.

Data Table 7 Site 1			
Test	Q-value	Weight	Total Q-value
Dissolved Oxygen		0.38	
pH		0.24	
TDS		0.16	
Temperature		0.22	

Overall Quality: _____

Data Table 8 Site 2			
Test	Q-value	Weight	Total Q-value
Dissolved Oxygen		0.38	
pH		0.24	
TDS		0.16	
Temperature		0.22	

Overall Quality: _____

Analyze and Conclude

1. **Analyzing Data** Using your data and your calculations, explain the overall water quality of your stream or lake.

_____

_____

_____

_____

_____

2. **Compare and Contrast** How did the measurements between Sites 1 and 2 compare? What might account for any differences?

_____

_____

_____

3. **Applying Concepts** Why would you expect the D.O. in a pond to be lower than in a rapidly moving stream? If applicable, did your measurements confirm this assumption? Explain.

_____

_____

_____

4. **Propose a Solution** Make recommendations to improve the water quality within your watershed.

_____

_____

_____

Table 9: Approximate Barometric Pressure at Different Elevations	
Elevation (m)	Pressure (mm Hg)
0	760
100	748
200	741
300	733
400	725
500	717
600	709
700	701
800	693
900	685
1000	676
1100	669
1200	661
1300	652
1400	643
1500	636
1600	628
1700	620
1800	612
1900	604
2000	596
2100	588
2200	580
2300	571

Table 10: 100% Dissolved Oxygen Capacity (mg/L)

	770 mm	760 mm	750 mm	740 mm	730 mm	720 mm	710 mm	700 mm	690 mm	680 mm	670 mm	660 mm
0°C	14.76	14.57	14.38	14.19	13.99	13.80	13.61	13.42	13.23	13.04	12.84	12.65
1°C	14.38	14.19	14.00	13.82	13.63	13.44	13.26	13.07	12.88	12.70	12.51	12.32
2°C	14.01	13.82	13.64	13.46	13.28	13.10	12.92	12.73	12.55	12.37	12.19	12.01
3°C	13.65	13.47	13.29	13.12	12.94	12.76	12.59	12.41	12.23	12.05	11.88	11.70
4°C	13.31	13.13	12.96	12.79	12.61	12.44	12.27	12.10	11.92	11.75	11.58	11.40
5°C	12.97	12.81	12.64	12.47	12.30	12.13	11.96	11.80	11.63	11.46	11.29	11.12
6°C	12.66	12.49	12.33	12.16	12.00	11.83	11.67	11.51	11.34	11.18	11.01	10.85
7°C	12.35	12.19	12.03	11.87	11.71	11.55	11.39	11.23	11.07	10.91	10.75	10.59
8°C	12.05	11.90	11.74	11.58	11.43	11.27	11.11	10.96	10.80	10.65	10.49	10.33
9°C	11.77	11.62	11.46	11.31	11.16	11.01	10.85	10.70	10.55	10.39	10.24	10.09
10°C	11.50	11.35	11.20	11.05	10.90	10.75	10.60	10.45	10.30	10.15	10.00	9.86
11°C	11.24	11.09	10.94	10.80	10.65	10.51	10.36	10.21	10.07	9.92	9.78	9.63
12°C	10.98	10.84	10.70	10.56	10.41	10.27	10.13	9.99	9.84	9.70	9.56	9.41
13°C	10.74	10.60	10.46	10.32	10.18	10.04	9.90	9.77	9.63	9.49	9.35	9.21
14°C	10.51	10.37	10.24	10.10	9.96	9.83	9.69	9.55	9.42	9.28	9.14	9.01
15°C	10.29	10.15	10.02	9.88	9.75	9.62	9.48	9.35	9.22	9.08	8.95	8.82
16°C	10.07	9.94	9.81	9.68	9.55	9.42	9.29	9.15	9.02	8.89	8.76	8.63
17°C	9.86	9.74	9.61	9.48	9.35	9.22	9.10	8.97	8.84	8.71	8.58	8.45
18°C	9.67	9.54	9.41	9.29	9.16	9.04	8.91	8.79	8.66	8.54	8.41	8.28
19°C	9.47	9.35	9.23	9.11	8.98	8.86	8.74	8.61	8.49	8.37	8.24	8.12
20°C	9.29	9.17	9.05	8.93	8.81	8.69	8.57	8.45	8.33	8.20	8.08	7.96
21°C	9.11	9.00	8.88	8.76	8.64	8.52	8.40	8.28	8.17	8.05	7.93	7.81
22°C	8.94	8.83	8.71	8.59	8.48	8.36	8.25	8.13	8.01	7.90	7.78	7.67
23°C	8.78	8.66	8.55	8.44	8.32	8.21	8.09	7.98	7.87	7.75	7.64	7.52
24°C	8.62	8.51	8.40	8.28	8.17	8.06	7.95	7.84	7.72	7.61	7.50	7.39
25°C	8.47	8.36	8.25	8.14	8.03	7.92	7.81	7.70	7.59	7.48	7.37	7.26
26°C	8.32	8.21	8.10	7.99	7.89	7.78	7.67	7.56	7.45	7.35	7.24	7.13
27°C	8.17	8.07	7.96	7.86	7.75	7.64	7.54	7.43	7.33	7.22	7.11	7.01
28°C	8.04	7.93	7.83	7.72	7.62	7.51	7.41	7.30	7.20	7.10	6.99	6.89
29°C	7.90	7.80	7.69	7.59	7.49	7.39	7.28	7.18	7.08	6.98	6.87	6.77
30°C	7.77	7.67	7.57	7.47	7.36	7.26	7.16	7.06	6.96	6.86	6.76	6.66

Lab 8 Comparing Respiration Rates in Seeds

Problem

What factors affect the rate of respiration in peas?

Introduction

Cellular respiration is the process that releases energy by breaking down glucose and other food molecules in the presence of oxygen. All organisms, including plants and animals, oxidize glucose to obtain energy. Often, this energy is used to convert ADP and phosphate into ATP. ATP is needed to carry out basic life functions such as growth. Plants carry out both the photosynthesis and cellular respiration. Seeds cannot photosynthesize, but they do contain a food supply that can be converted to ATP. This ATP allows the embryo to grow and develop when conditions are right. However, seeds only carry out cellular respiration. What factors affect the rate of respiration in seeds?

In this lab, you will test the rate of respiration by measuring the levels of carbon dioxide under two conditions. You will explore the effects of temperature on germination and the differences between germinating and nongerminating seeds.

Skills Focus

Observe, Predict, Use a Data-Collection Interface and Sensor

Materials

- Vernier data-collection interface[1]
- Vernier CO_2 Gas Sensor
- 250-mL respiration chamber
- 25 germinating peas
- 25 nongerminating peas
- ice cubes
- paper towel
- thermometer
- 100-mL beakers (2)
- Logger *Pro* (optional)

Safety

Always wear goggles. Wash your hands thoroughly with soap and warm water at the end of the lab.

Pre-Lab Questions

1. **Recall** What is the equation for cellular respiration?

   _____

2. **Relate Cause and Effect** Explain why it is possible to compare the rates of respiration by measuring carbon dioxide levels.

   _____

   _____

   _____

3. **Predict** Which peas do you think will have the highest rate of respiration? Explain your answer.

   _____

   _____

   _____

   _____

   _____

Procedure

1. Put on goggles.

2. Measure the room's temperature using a thermometer and record it in the Temperature column of the Data Table.

3. Connect the CO_2 Pressure Sensor to your data-collection interface. Prepare the interface for data collection. Set the collection rate to 0.2 samples/second and the data-collection length to 300 seconds.[10]

4. Blot 25 germinating peas between two pieces of paper towel. Place the germinating peas into the respiration chamber. Place the shaft of the CO_2 Gas Sensor in the opening of the respiration chamber as shown in Figure 1. Wait 1 minute, and then start data collection.[2] Data will be collected for 5 minutes.

5. Save data.[3]

6. Remove the CO_2 Gas Sensor from the respiration chamber. Place the peas in a 100-mL beaker filled with cold water and ice cubes. Allow the peas to soak in cold water for 10 minutes. While they soak, continue to Step 7.

Figure 1 Completed setup

7. Use a notebook to fan air across the openings in the shaft of the CO_2 Gas Sensor for 1 minute. Fill the respiration chamber with water and then empty it. Thoroughly dry the inside of the respiration chamber with a paper towel.

8. Repeat Steps 2–7, but use nongerminating peas instead of germinating peas. In Step 6, place the nongerminating peas on a dry paper towel and not in the ice bath.

9. Measure the temperature of the cold water with a thermometer and record it in the Temperature column of the Data Table.

10. Remove the peas from the cold water and repeat Steps 4–6 using the cold peas.

11. Clean up as directed by your teacher.

Organize Data

11. Use the data-collection software to graph all three runs of data on a single graph.[5]

12. Find the rate of respiration for each run, by finding the slope, m.[4] Record the rate of respiration in the Data Table.

Data Table		
Peas	Temperature (°C)	Rate of Respiration (ppm/min)
Germinating		
Nongerminating		
Germinating		

Analyze and Conclude

1. **Interpret Graphs** What evidence is there that cellular respiration has occurred?

_____

_____

2. **Interpret Data** What is the effect of germination on the rate of cellular respiration in peas?

_____

_____

_____

_____

3. **Interpret Data** What is the effect of temperature on the rate of cellular respiration in peas?

_____

_____

_____

4. **Apply Concepts** Why do germinating peas undergo cellular respiration?

_____

_____

_____

5. **Predict** How would the rate of cellular respiration change if you add water to the dry peas, keeping them at room temperature?

_____

_____

_____

Extend Your Inquiry

Is respiration equal between different organisms? Design an experiment to test the rate of respiration in different organisms. You could test germination of different types of seeds or use a small animal such as insects or earthworms. **IMPORTANT:** If your experimental design involves the use of animals be sure to treat them with care and avoid injuring them or you.

Lab 9 Transpiration Rates

Problem

What factors affect the rate of transpiration in plants?

Introduction

Water transport is an essential process in land plants. Water is obtained by roots at the bottom of the plant but is used by leaves at the top. Efficiently transporting water between the roots and leaves is needed for the plant's survival. In leaves, the walls of mesophyll cells are kept moist so that gases can enter and leave the cells easily. As a result, water continuously evaporates from the surface of leaves in a process known as transpiration. As each water molecule evaporates, cohesion pulls more water molecules out from the xylem vessels in the leaf. This pull is transmitted all the way down from the leaves to the root in a continuous water column, allowing the efficient movement of water from the roots up to the leaves. Any factor that affects a plant's transpiration rate affects water transport throughout the plant. This lab will measure transpiration rates under different conditions of light, humidity, temperature, and air movement. The data will be collected by measuring pressure changes as the plant takes up water into the stem.

Skills Focus

Observe, Predict, Use a Data-Collection Interface and Sensor

Materials

- Vernier data-collection interface[1]
- Vernier Gas Pressure Sensor, with provided plastic tube
- 3 utility clamps
- ring stand
- tomato seedlings
- plastic tube clamp
- dropper or Beral pipette
- scalpel
- thermometer
- metric ruler
- masking tape
- 100-watt light source
- plastic gallon sized bag with twist tie
- heater, small electric
- fan with slow speed
- plant mister
- plastic syringe
- shallow bowl, with hard walls

Safety 🥽 ✂️ 🧼

Always wear goggles. Do not direct the point of the scalpel toward yourself or others. Use the scalpel only as instructed. Wash your hands thoroughly with soap and warm water at the end of the lab.

Pre-Lab Questions

1. **Recall** Which specialized cells regulate the movement of gases into and out of leaf tissue?

   _____

2. **Apply Concepts** In Step 3 of the lab, why is it important to submerge the plant under water?

   _____

   _____

   _____

3. **Predict** Which environmental condition do you think will have the greatest effect on the rate of transpiration?

   _____

   _____

   _____

Procedure

Part A: Setting Up the Control

1. Position the ring stand, utility clamps, and Gas Pressure Sensor as shown in Figure 1. Do not place the plastic tube or the plant in the apparatus yet.

Figure 1 Completed setup

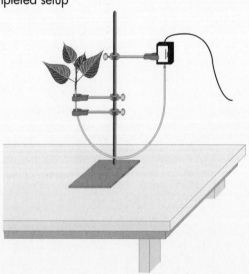

2. Place the tip of the plastic syringe into the end of the plastic tube without the white plastic connector. Submerge the other end of the tube into water. Use the syringe to draw water up into the tube until it is full. Check the tube closely for air bubbles. If you see any, you may try to expel them by tapping the tube. If this does not work, you may have to start the process of filling the tube over again.

3. Slip a plastic tube clamp on the tube as shown in Figure 2. Bend the tube into a U shape with both ends up, being careful to hold both ends at equal heights. If one end is lower than the other, water will drip out. With the tube full of water, remove the syringe.

Figure 2 Tube clamp on tube

4. Select a plant that has a stem about the same diameter as the opening of the tube. Use a scalpel to carefully cut the plant one inch above the soil. Submerge the plant under water in a shallow bowl. Against the bottom of the bowl, carefully make a cut at a 45° angle near the base of the stem.

5. Raise the end of the tube with the white plastic connector until water begins to drip out of the other end. Carefully push the cut stem of the plant into the end of the tube where the water is dripping out. Be careful not to allow any air bubbles to form between the cut portion of the stem and the water in the tube. Push the plant at least 1 centimeter into the tube. Be careful not to damage the plant. Shut the tube clamp as tightly as possible, as shown in Figure 3.

Figure 3 Clamping plant in tube

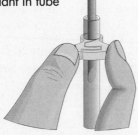

6. When the tube clamp is shut tight, invert your plant cutting to check for any leaks. If water does leak out, turn the plant right side up and try tightening the clamp further. **IMPORTANT**: Be sure the tube is completely filled with water. The water column must be flush with the stem. There should be no air visible at the base of the stem. If water moves down the tube away from the stem after it has been inserted, check for a leak in the system.

7. Connect the opposite end of the plastic tube to the sensor valve. **CAUTION:** Do not allow water to enter the valve of the sensor.

8. Secure the plant in an upright position with the utility clamps as shown in Figure 1. It should be positioned so that the cut stem is about 8 cm below the water level at the other end of the tube, as shown in Figure 1. On the end of the tube without the plant, place a mark at the water level to allow you to refill the tube to the starting level in Step 18.

9. Review your setup, and check it carefully against the diagram in Figure 1. The water column must be continuous for the experiment to work. If any air bubbles have formed, you may need to start over.

10. Place your plant setup in an area where the wind, humidity, and temperature are reasonably constant. This location will be your control setup. If you move your plant, be sure to hold the ends of the tubing steady, so they do not move up or down relative to each other.

11. Allow the system 5 minutes to adjust to the environment. While the system is adjusting, set up your data-collection interface.

12. Connect the Gas Pressure Sensor to the interface.[11] Set the data-collection length to 900 seconds. Set the data collection rate to 0.1 samples per second.

13. Check the base of the plant stem in the water tube to make sure that no air bubbles or air pockets have formed that will prevent the plant from taking up water. If an air pocket has formed, refit the plant in the tube before initiating data collection in Step 14.

14. Start data collection.[2]

15. After 15 minutes, find the value of the slope, m, for the change in pressure.[4]

Part B: Design Your Own Experiment

16. As a class, you will test the effect of a variety of environmental factors on transpiration. Each group will be assigned by your teacher to test the effect of one of the following environmental factors: light intensity, wind, humidity, temperature.

17. Design an experiment to simulate an increase or decrease in the environmental factor you are assigned. Check your procedure with your teacher and obtain the materials needed for the experiment.

18. Refill the water level in the tube to the same level marked in Step 8. Be sure there are no air bubbles, and then begin your data collections. Record your values in Data Table 1.

Part C: Determine the Rate of Transpiration

1. Cut all the leaves (not stems) off your plant. Put all of the leaves on a balance and measure the mass. Record the value in Data Table 1.

2. To find the surface area, cut out a section of leaf 5 cm × 5 cm. Determine the mass for this leaf section and divide by 25 cm² to find the mass of l cm² of leaf. Divide the total mass of the leaves by the mass of l cm² to find the total leaf surface area. Record the calculated surface area in Data Table 1.

3. Calculate the rate of transpiration. To do this, divide the slope by the surface area for each plant. These rate values can be expressed as kPa/min/cm². Record the rate of transpiration in Data Table 1.

4. Find the adjusted rate for your simulated condition by subtracting the rate of transpiration of the control from the rate of transpiration of the simulated condition. Record this adjusted rate in the last column of Data Table 1.

Data Table 1					
Test	Slope (kPa/min)	Mass (g)	Surface Area (cm²)	Rate/Area (kPa/min/cm²)	Adjusted Rate (kPa/min/cm²)
Simulated environmental condition					
Control					

5. Share the adjusted rate for your environmental condition with the class. Record the class results in Data Table 2 for each of the tested environmental conditions. If a condition was tested by more than one group, use the average value.

Data Table 2	
Class Data	
Simulated Environmental Condition	Adjusted Rate (kPa/min/cm²)
Light	
Humidity	
Wind	
Temperature	

6. **Graph** Make a bar graph to show the effect of different environmental conditions on the transpiration in plant cuttings. Use the data in Data Table 2 to plot the adjusted rate for each test on the y-axis and the condition on the x-axis.

Analyze and Conclude

1. **Interpret Graphs** Which simulated environmental condition resulted in the greatest rate of water loss? Explain why this factor might increase water loss when compared to the others.

_____

_____

_____

_____

2. **Draw Conclusions** Under which of the tested environmental conditions would you expect to see more closed stomata? Use your data to explain your answer.

_____

_____

_____

_____

3. **Applying Concepts** Why is it necessary to determine the surface area of the leaves in order to calculate the rate of transpiration?

_____

_____

_____

_____

4. **Connect Concepts** Scientists sometimes measure transpiration rates and use them to figure out how much photosynthesis is occurring. Explain how this works.

_____

_____

_____

_____

Extend Your Inquiry

Test the rate of transpiration in different types of plants. Set up a similar experiment, but use two or three different types of plants.

Lab 10 **Human Respiration**

Problem

What affects the rate of respiration in humans?

Introduction

Through the process of breathing, your body obtains oxygen, releases carbon dioxide, and regulates the blood's pH level. The process of taking in air is known as *inhalation,* while the process of blowing out air is called *exhalation.* The rate of respiration depends on the levels of oxygen and carbon dioxide in your blood.

In this lab, you will monitor the respiratory patterns of one member of your group under different conditions. A respiration belt will be strapped around the test subject and connected to a data-collection interfaced Gas Pressure Sensor. Each respiratory cycle, one inhalation and one exhalation, will be recorded by the interface, allowing you to calculate and compare respiratory rates under different conditions.

Skills Focus

Use Sensors and a Data-Collection Interface, Analyze Data, Infer

Materials

- Vernier data-collection interface[1]
- Vernier Gas Pressure Sensor
- Vernier Respiration Monitor Belt
- respiration bag

Safety ⚠

If you are prone to dizziness or nausea, do not volunteer to be the test subject. If the person being tested experiences dizziness, nausea, or a headache during testing, stop data collection and inform your teacher.

Pre-Lab Questions

1. **Use Models** In Part C of this lab, what type of situation does rebreathing exhaled air simulate?

_____

_____

_____

2. **Predict** Under which condition will the test subject's rate of respiration be greater?

_____

_____

Procedure

Part A: Set Up the Experiment

1. Select a member of your lab group as the test subject. Wrap the Respiration Monitor Belt snugly around the subject's rib cage. Press the hook and loop fastener strips together at the back. Position the belt so the air bladder is resting over the diaphragm as shown in Figure 1.

Figure 1 Completed Setup

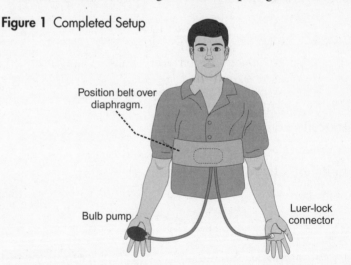

Position belt over diaphragm.

Bulb pump

Luer-lock connector

2. Attach the Respiration Monitor Belt to the Gas Pressure Sensor. There are two rubber tubes connected to the bladder. One tube has a white Luer-lock connector at the end and the other tube has a bulb pump attached. Connect the Luer-lock connector to the stem on the Gas Pressure Sensor with a gentle half turn.

3. Prepare data-collection interface for data collection. Set the data-collection rate to 10 samples/second and the data-collection length to 180 seconds.[12]

4. Have the test subject sit upright in a chair. Close the shut-off screw of the bulb pump by turning it clockwise as far as it will go. Pump air into the bladder by squeezing the bulb pump. Fill the bladder as full as possible without making it uncomfortable for the test subject.

Part B: Holding of Breath

5. Start data collection. For the first 30 seconds, the test subject should breathe normally. When 30 seconds have passed, the test subject should take a deep breath and hold it as long as he or she can without experiencing extreme discomfort or concern. After the subject releases his or her breath, instruct him or her to breathe normally for the remainder of data collection. Data will be collected for 3 minutes.

6. When data collection has finished, examine the graph on your interface.[6] Determine the respiration rate before and after the test subject's breath was held and record these rates in Data Table 1.

Part C: Breathing Exhaled Air

7. Prepare data-collection interface for a new data collection.[12] Change the data-collection rate to 2 samples/second and the data-collection length to 60 seconds.

8. Obtain a respiration bag from your teacher. The subject will breathe in and out of the bag through the attached cup during this portion of the experiment.

9. The subject will breathe in and out of the respiration bag for a period of 6 minutes. During the 6-minute period, three series of data will be collected. Use a watch or clock to make sure that data collection is performed at the proper times. **IMPORTANT:** If the test subject experiences dizziness, nausea, or a headache during data collection, testing should be stopped immediately and the subject should remove the respiration bag so he or she can breathe normally.

10. Fill the respiration bag with air and have the subject hold the breathing cup snugly to his or her face so that outside air does not get in.

11. Have the subject begin breathing into and out of the bag through the attached cup. Check to make sure that he or she is able to easily breathe both in and out of the respiration bag. The subject must breathe from the bag for the entire 6 minutes. Start data collection. Data will be collected for 1 minute. During that time, look ahead to Steps 12 and 13 and prepare to follow the instructions.

12. When 1 minute has elapsed, data collection will stop.

    a. One member should perform Step 13 **immediately.**

    b. Another member of your lab group must continue to keep track of the total elapsed time.

    c. The subject must continue breathing into the bag for the entire 6-minute period.

13. Calculate the relative amplitude of the respiration waves for the data collection. You must complete this step within 1 minute.

   a. Follow the instructions for your data-collection interface to find specific statistics for the data on your graph.[7]

   b. Subtract the minimum pressure value from the maximum pressure value (in kPa) to calculate the amplitude of the respiration waveform.

   c. Record the calculated wave amplitude for this time interval ("0 to 1 Minute") in Data Table 2.

14. After a total of 2.5 minutes have elapsed, with the subject still breathing into the bag, continue the data collection. Repeat Steps 12–13, and record the amplitude value in Data Table 2 under "2.5 to 3.5 Minutes."

15. After a total of 5 minutes have elapsed, with the subject still breathing into the bag, start data collection. Repeat Steps 12–13, and record the amplitude value in Data Table 2 for "5 to 6 Minutes."

Data Table 1	
Holding of Breath: Respiration Rates	
Before Holding Breath	After Holding Breath
_____ breaths/minute	_____ breaths/minute

Data Table 2		
Breathing Exhaled Air: Amplitudes of Respiration Waves		
0 to 1 Minute	2.5 to 3.5 Minutes	5 to 6 Minutes
_____ kPa	_____ kPa	_____ kPa

Analyze and Conclude

1. **Analyze Data** Did the rate of respiration of the test subject change after holding his or her breath? If so, describe how it changed.

   _____

   _____

   _____

2. **Draw Conclusions** What is the significance of an increase in the amplitude and frequency of the waveform while the test subject was breathing into the bag?

   _____

   _____

   _____

3. **Interpret Data** How did the respiratory waveforms change while the test subject was breathing exhaled air from the bag? How would you interpret this result?

   _____

   _____

   _____

   _____

4. **Apply Concepts** Based on your results, how does carbon dioxide affect your breathing?

   _____

   _____

   _____

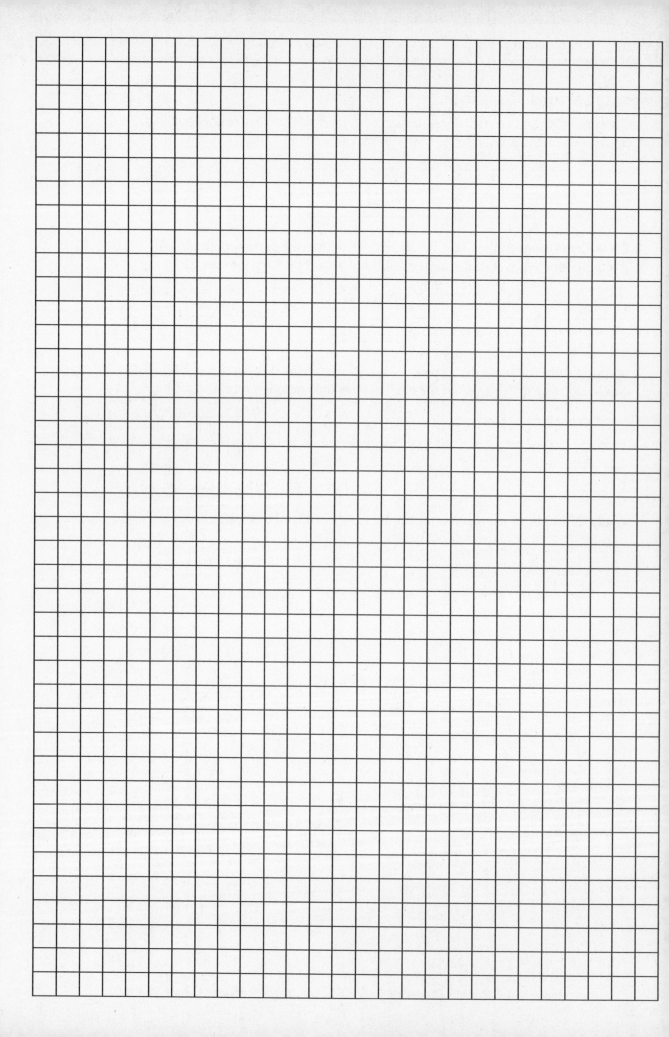

Texas Instruments Labs

Teacher Notes and Answers are on the CD-ROM.

The lab activities from Texas Instruments are listed in the table below along with the sensor required to conduct each lab.

Lab Activity	Probeware
Enzyme Action	TI-83/84 Calculator, Vernier O_2 Gas Sensor
Testing Watersheds	TI-83/84 Calculator, Vernier Dissolved O_2 Sensor, Vernier Conductivity Sensor, Vernier Temperature Sensor
Comparing Respiration Rates in Seeds	TI-83/84 Calculator, Vernier CO_2 Gas Sensor
Transpiration Rates	TI-83/84 Calculator, Vernier Gas Pressure Sensor
Human Respiration	TI-83/84 Calculator, Vernier Respiration Belt

The CD-ROM provided in the front of the book contains both a PDF and an editable copy of the student version of each lab. In addition, the CD-ROM includes a teacher's version with an answer guide and teaching tips. Finally, the CD-ROM includes information about the use of individual data-collection systems and sensors specific to each lab.

These labs have been adapted from Experiments 6A Enzyme Action, 20 Watershed Testing, 11B Cell Respiration, 10 Transpiration, and 26 Control of Human Respiration from the *Biology with Vernier Lab Manual* by Kelly Redding and David Masterman.

Lab 11 **Enzyme Action**

Problem

How do concentration and temperature affect the rate of an enzyme-catalyzed reaction?

Introduction

Enzymes act as catalysts, increasing chemical reaction time in living organisms. For an enzyme to function properly, temperature is very important. Most organisms have a specific temperature range in which they survive, and their enzymes function best within that temperature range. Outside of that range, enzymes may cease to function.

Hydrogen peroxide (H_2O_2) is toxic to most living organisms. Many organisms decompose hydrogen peroxide with enzymes before it can do much damage, as it can be broken down into two harmless products—water (H_2O) and oxygen (O_2). Although this reaction occurs spontaneously, enzymes increase the rate of reaction. The enzyme catalase helps increase the rate at which hydrogen peroxide is broken down.

In this lab, you will measure the rate of enzyme activity under two conditions: different enzyme concentrations and at different temperatures.

Skills Focus

Use Sensors and a Data-Collection Interface, Analyze Data, Infer

Materials

- TI-83 Plus or TI-84 Plus graphing calculator
- EasyData application
- Vernier data-collection interface
- Vernier O_2 Gas Sensor
- 400-mL beaker
- 10-mL graduated cylinder
- 18 × 150 mm test tubes (3)
- 250-mL Nalgene bottle
- 3.0% hydrogen peroxide solution
- yeast suspension
- ice
- test tube rack
- thermometer
- 3 dropper pipettes
- hot plate
- Logger *Pro* (optional)

Safety 🗿 🧍 ⚗️ 🧪 ⚡ 🧤 🔬

Always wear goggles and an apron when using a water bath. If you have glass beakers or test tubes, check for cracks or chips. Alert your teacher if you break a glass object. Avoid contact with the surface of the hot plate. To avoid electric shocks, make sure cords, connections, and your hands are dry. Wash your hands thoroughly with soap and warm water at the end of the lab.

Pre-Lab Questions

1. **Relate Cause and Effect** How will you know that a chemical reaction is taking place in Part B and Part C?

_____

_____

2. **Predict** Which concentration of yeast suspension do you think will increase the reaction time the most? Why?

_____

_____

_____

3. **Relate Cause and Effect** Why are you measuring oxygen levels?

_____

_____

_____

Procedure

Part A: Set Up the Experiment

1. Obtain and wear goggles.
2. Turn on the calculator. Connect the O_2 Gas Sensor, data-collection interface, and calculator.
3. Set up the data-collection mode.
   a. Start the EasyData application if it is not already running.
   b. Select 〔File〕 from the Main screen, and then select **New** to reset the application.
   c. Select 〔Setup〕 from the Main screen, and then select **Time Graph...**
   d. Select 〔Edit〕 on the Time Graph Settings screen.
   e. Enter 5 as the time between samples in seconds, and select 〔Next〕.
   f. Enter 36 as the number of samples, and select 〔Next〕 (data will be collected for 3 minutes).
   g. Select 〔OK〕 to return to the Main screen.

4. Label three test tubes 1, 2, and 3. Fill each test tube with 5 mL of 3.0% H_2O_2 and 5 mL of water. Place each test tube on the test tube rack.

Part B: Testing the Effect of Enzyme Concentration

5. Using a clean dropper pipette, add 5 drops of yeast suspension to test tube 1. Cover the opening of the test tube with a gloved finger and gently invert the test tube two times. Pour the contents of the test tube into a clean 250-mL Nalgene bottle. Place the O_2 Gas Sensor into the bottle as shown in Figure 1. Gently push the sensor down into the bottle until it stops. The sensor is designed to seal the bottle with minimal force. After 30 seconds, select ⌈Start⌉ to begin data collection.

Figure 1 Completed Setup

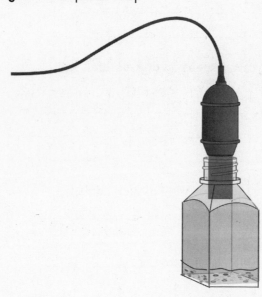

6. When data collection has finished, a graph of oxygen gas versus time will be displayed.

7. Remove the O_2 Gas Sensor from the Nalgene bottle. Rinse the bottle with water and dry it with a paper towel.

8. Find the value of the slope, *a,* for the change in oxygen concentration. The slope is the rate of the reaction. In your data table, record the value of the slope, *a.* To find the slope:

   a. Select ⌈Anlyz⌉, and then select **Linear Fit.**

   b. The linear-regression statistics for these two lists are displayed for the equation in the form:

   $$y = ax + b$$

   c. Enter the absolute value of the slope, *a,* as the reaction rate in Data Table 2.

   d. Select ⌈OK⌉ to view a graph of the data and the regression line.

   e. Select ⌈Main⌉ to return to the Main screen.

9. Store the data from the first run so that it can be used later.

   Select (File), and then select **Store Run.**

   Select (OK) to store your latest data and overwrite the data in Lists 3 and 4 (L3 and L4).

10. Repeat Steps 5–9, adding 10 drops of yeast suspension to test tube 2. **NOTE:** After selecting (Start) to begin data collection, select (OK) to start collecting data. Your stored data will not be overwritten.

11. Repeat Steps 5–9, adding 20 drops of yeast suspension to test tube 3.

12. Graph all three runs of data on a single graph. To do this:

    a. From the Main screen, select (Graph).

    b. Select (Adv), and then select **L2, L3 and L4 vs L1.**

    c. All three runs should now be displayed on the same graph. Each point of the 5-drop run is plotted with a plus sign, each point of the 10-drop run is plotted with a square, and each point of the 20-drop run is plotted without a marker.

    d. When finished with the graph, select (Main) to return to the Main screen.

Data Table 1		
Test Tube	Yeast Concentration	Slope, or Rate of Reaction (%/s)
1	5 drops	
2	10 drops	
3	20 drops	

Part C: Testing the Effect of Temperature

As a class, you will test the effects of temperature on the enzyme. Each group will test a specific temperature range and share its data with the rest of the class. Using your assigned temperature range, set up your water bath as described below. Place a thermometer in your water bath to assist in maintaining the proper temperature.

- 0–5°C: Fill a 400-mL beaker with ice and water.

- 20–25°C (room temperature): No water bath is needed.

- 30–35°C: Fill a 400-mL beaker with very warm water; heat it on a hot plate.

- 50–55°C: Fill a 400-mL beaker with hot water; heat it on a hot plate.

13. Rinse the three numbered test tubes used for Part I. Fill each test tube with 5 mL of 3.0% H_2O_2 and 5 mL of water, and then place the test tubes in the water bath. The test tubes should be in the water bath for 5 minutes before proceeding to Step 15. Record the temperature of the water bath, as indicated on the thermometer, in the space provided in Data Table 2.

14. Repeat Step 2.

15. Add 10 drops of the enzyme solution to each test tube. For each test tube repeat Steps 5–9. **NOTE:** After selecting ⌈Start⌉ to begin data collection, select ⌈OK⌉ to start collecting data.

16. Record the reaction rates in Data Table 2. Calculate the average reaction rate for the three trials you tested. Record the average rate in Data Table 2.

Data Table 2	
Test Tube Label	**Slope, or Rate (%/s)**
Trial 1	
Trial 2	
Trial 3	
Average	
Temperature range: °C– °C	

17. Record the average rate and the temperature of your water bath from Data Table 2 on the class chalkboard. When the entire class has reported its data on the chalkboard, record the class data in Data Table 3.

Data Table 3 (Class Data)		
Temperature Range	**Actual Temperature**	**Average Rate of Reaction (%/s)**
0–5°C range		
20–25°C range		
30–35°C range		
50–55 °C range		

18. **Graph** Using the data from Data Table 3, make a graph of the rate of enzyme activity versus temperature by hand or using Logger *Pro* software. Plot the rate values from the class data on the *y*-axis and the temperature on the *x*-axis.

19. Follow your teacher's instructions for dismantling the setup and for cleanup. Wash your hands thoroughly with soap and warm water before leaving the lab.

Analyze and Conclude

1. **Interpret Tables** Refer to Data Table 1. Which concentration of yeast had the highest reaction rate? Which concentration of yeast had the lowest rate?

_____

_____

_____

2. **Predict** If you increased the concentration of the enzyme solution to 30 drops, what do you think will happen to the rate of reaction?

_____

_____

_____

3. **Interpret Graphs** Refer to your graph. At what temperature is the rate of enzyme activity the highest? Lowest? What does this indicate about enzyme activity?

_____

_____

_____

4. **Apply Concepts** Why is it important for your body to maintain a stable internal temperature?

_____

_____

_____

Extend Your Inquiry

How can pH affect catalase activity? Design an experiment to test your hypothesis. Ask your teacher for comments on your design. If necessary, revise the procedure. With your teacher's permission and supervision, carry out your lab.

Lab 12 **Testing Watersheds**

Problem
How do you determine water quality?

Introduction
On a hot summer day at Arizona's Slip Rock State Park, a sign states "Unsafe water, no swimming." Every day at the park the water quality is tested to ensure that bacteria levels are not at a harmful level. In addition to knowing if it is safe to swim, there are many reasons for determining water quality. We depend on fresh water for drinking and growing crops. Pollutants in the water can affect an entire watershed.

How is water quality measured? In the 1970s, the National Sanitation Foundation developed a standard index for measuring water quality. The Water Quality Index, or WQI, is a series of nine tests that measures temperature, pH, turbidity, total solids, dissolved oxygen (D.O.), biochemical oxygen demand, phosphates, nitrates, and fecal coliform. Through these nine tests, the water quality value, or Q-value, of the water is determined.

In this lab, you will conduct a modified version of the WQI tests. You will test only water temperature, dissolved oxygen, pH, and total dissolved solids. This modified version of the WQI will allow you to determine the general quality of the stream or lake you are sampling.

Materials
- TI-83 Plus or TI-84 Plus graphing calculator
- EasyData application
- Vernier data-collection interface
- Vernier Dissolved Oxygen Sensor, with Dissovled Oxygen Filling Solution
- Vernier pH Sensor
- Vernier Conductivity Sensor
- Vernier Temperature Sensor
- 4 water sampling bottles
- large plastic cup or beaker
- D.O. calibration bottle
- D.O. filling solution
- distilled water
- pipette
- permanent marker

Skills Focus
Measure, Analyze Data, Use a Calculator, Data-Collection Interface and Sensors

Safety 🔥 ⚠️ 🧤
Be aware of dangerous plants that are common in your area. Be careful in and near the water. Do not enter any fast-moving streams. After performing the lab, wash your hands with warm water.

Pre-Lab Questions

1. **Apply Concepts** In Step 7, why is it important to keep the bottle under water when you add the stopper?

_____

_____

_____

2. **Infer** Think about the area where you will be testing water quality. Are there any sources of pollution that you know of? How might pollution affect your results?

_____

_____

_____

Procedure

Part A: Setting Up Dissolved Oxygen Sensor and Calculator

1. To prepare the Dissolved Oxygen Sensor, remove the protective cap. Unscrew the membrane cap from the tip of the sensor. Using a pipette, fill the membrane cap with 1 mL of D.O. Electrode Filling Solution. Carefully thread the membrane cap back onto the electrode. Place the sensor into a 250-mL beaker containing distilled water.

2. Turn on the calculator and connect it to the data-collection interface. Connect the Dissolved Oxygen Sensor to Channel 1, connect the pH Sensor to Channel 2, and connect the Temperature Sensor to Channel 3.

3. Start the EasyData application, if it is not already running. Select ⌈File⌉ from the Main screen, and then select **New** to reset the application.

4. Prior to taking any readings, allow the Dissolved Oxygen sensor to warm up. With the sensor still in the distilled water, wait 10 minutes. The sensor must stay connected at all times to function properly.

5. Calibrate the Dissolved Oxygen Sensor according to your teacher's instruction.

Part B: Collecting Samples

You will take four water samples to measure the dissolved oxygen concentration, pH, water temperature, and total dissolved solid (TDS) concentration at two locations. Your teacher will select two sites at which you will make measurements.

6. Carefully enter the water and move away from the edge as long as it remains safe. Be careful of slippery and sharp rocks or fast-moving water. **CAUTION:** If the current is strong, do not enter the water.

7. Rinse the sampling bottle several times with stream water. Place the bottle below the surface and allow water to flow into the opening for 3 minutes. Fill the bottle so it is completely full, and place the stopper

in the bottle while the bottle is still under water. Label the sample bottle Site 1a.

8. Take your sample away from the water to make measurements.

9. Remove the Dissolved Oxygen Sensor from its storage bottle. Place the sensor into the water and gently swirl to allow water to move past the sensor's tip. **IMPORTANT:** Place only one sensor in the water at a time. When the reading is stable, record the dissolved oxygen concentration in Data Table 1. Remove the sensor from the water and place it back into the storage bottle.

10. Place the Temperature Sensor into the water sample. When the reading on the calculator screen has stabilized, record the temperature in Data Table 1. Remove the sensor from the water.

11. Remove the pH Sensor from its storage bottle. Place the sensor into the water sample. When the reading has stabilized, record the pH in Data Table 1. Remove the sensor from the water and place it back into the storage bottle.

12. Cover the sampling bottle and take it with you. You will need it for Part C of the lab.

13. Choose a spot 6 meters from Site 1a. Repeat Steps 6–10. Label it as Site 1b.

14. Repeat Steps 6–13 at a location at least 1 km away from Site 1. Label the sampling bottles as Site 2a and Site 2b.

Part C: Measuring Total Dissolved Solids (TDS)

15. Set the switch on the Conductivity Sensor box to 0–2000 μS (2000 μS = 1000 mg/L TDS).

16. Unplug all sensors from the data-collection interface. Plug the Conductivity Sensor into Channel 1 of the interface.

17. Start the EasyData application, if it is not already running. Select (File) from the Main screen, and then select **New** to reset the application. Select (Setup), and then select **Single Point**.

18. Calibrate the Conductivity Sensor according to your teacher's instruction.

19. To find the Total Dissolved Solids (TDS) in the water, place the tip of the conductivity sensor into the sample water from Site 1a. The hole near the tip of the sensor should be completely submerged. When the reading has stabilized, select (Start) to begin sampling. **IMPORTANT:** Leave the sensor tip submerged while data is being collected for the next 10 seconds. After 10 seconds, the TDS value will appear on the calculator screen. Record this value in Data Table 1 (round to the nearest 1 mg/L TDS). Select (OK) to return to the Main screen. When finished, rinse the sensor with distilled water.

20. Repeat Step 19 for the water samples collected from the other three sites.

Part D: Collect Data and Determine Water Quality Index

21. Calculate the averages for measurements at Sites 1 and 2, and record the results in Data Table 1.

Data Table 1				
Location	Temperature (°C)	Dissolved Oxygen (mg/L)	pH	Total Dissolved Solids (mg/L)
Site 1a				
Site 1b				
Average				
Site 2a				
Site 2b				
Average				
Temperature Difference				

22. Record the average levels of dissolved oxygen that you measured at each site in Data Table 2.

23. Determine the barometric pressure in mm Hg by using either a barometer or Table 9: Approximate Barometric Pressure at Different Elevations.

24. Use barometric pressure and temperature values to determine the dissolved oxygen saturation. Look up the level of dissolved oxygen for air-saturated water in mg/L from Table 10: 100% Dissolved Oxygen Capacity. Record the value for each site in Data Table 2.

25. Determine the % saturation by using the following formula:

$$\% \text{ saturation} = \frac{\text{Measured D.O. level}}{\text{Saturated D.O. level}} \times 100$$

Record the % saturation of dissolved oxygen in Data Table 2.

Data Table 2 Dissolved Oxygen (% saturated)			
	Average Dissolved Oxygen (mg/L)	Dissolved Oxygen in Saturated Water	% Saturated
Site 1			
Site 2			

26. Using Tables 3–5, determine the Q-value for dissolved oxygen, pH, and TDS. Use the number closest to your value to obtain the closest Q-value. Record your results in Data Table 7 for Site 1 and in Data Table 8 for Site 2.

Table 3	
Dissolved Oxygen (D.O.) Test Results	
D.O. (% saturation)	Q-value
0	0
10	5
20	12
30	20
40	30
50	45
60	57
70	75
80	85
90	95
100	100
110	95
120	90
130	85
140	80

Table 4	
pH Test Results	
pH	Q-value
2.0	0
2.5	1
3.0	3
3.5	5
4.0	8
4.5	15
5.0	25
5.5	40
6.0	54
6.5	75
7.0	88
7.5	95
8.0	85
8.5	65
9.0	48
9.5	30
10.0	20
10.5	12
11.0	8
11.5	4
12.0	2

27. Find the difference between the average temperatures of Sites 1 and 2. Record the difference in Data Table 1.

28. Use Data Table 6 and the temperature difference to determine the Q-value for temperature. Record your result in both Data Table 7 and Data Table 8.

29. Multiply each Q-value by the given weighting factor. Record the total Q-value in Data Tables 7 and 8.

Data Table 5	
Total Dissolved Solids (TDS) Test Results	
TDS (mg/L)	Q-value
0	80
50	90
100	85
150	78
200	72
250	65
300	60
350	52
400	46
450	40
500	30

Data Table 6	
Temperature Test Results	
Temperature Difference (°C)	Q-value
0	95
5	75
10	45
15	30
20	20
25	15
30	10

30. Determine the overall water quality of your stream by adding the four total Q-values in Table 7 for Site 1 and in Table 8 for Site 2. Record the result. The closer this value is to 100, the better the water quality.

Data Table 7 Site 1			
Test	Q-value	Weight	Total Q-value
Dissolved Oxygen		0.38	
pH		0.24	
TDS		0.16	
Temperature		0.22	

Overall Quality: _____

Data Table 8 Site 2			
Test	Q-value	Weight	Total Q-value
Dissolved Oxygen		0.38	
pH		0.24	
TDS		0.16	
Temperature		0.22	

Overall Quality: _____

Analyze and Conclude

1. **Analyzing Data** Using your data and your calculations explain, the overall water quality of your stream or lake.

_____

_____

_____

_____

_____

2. **Compare and Contrast** How did the measurements between Sites 1 and 2 compare? What might account for any differences?

_____

_____

_____

3. **Applying Concepts** Why would you expect the D.O. in a pond to be lower than in a rapidly moving stream? If applicable, did your measurements confirm this assumption? Explain.

_____

_____

_____

4. **Propose a Solution** Make recommendations to improve the water quality within your watershed.

_____

_____

_____

Table 9: Approximate Barometric Pressure at Different Elevations	
Elevation (m)	Pressure (mm Hg)
0	760
100	748
200	741
300	733
400	725
500	717
600	709
700	701
800	693
900	685
1000	676
1100	669
1200	661
1300	652
1400	643
1500	636
1600	628
1700	620
1800	612
1900	604
2000	596
2100	588
2200	580
2300	571

Table 10: 100% Dissolved Oxygen Capacity (mg/L)

	770 mm	760 mm	750 mm	740 mm	730 mm	720 mm	710 mm	700 mm	690 mm	680 mm	670 mm	660 mm
0°C	14.76	14.57	14.38	14.19	13.99	13.80	13.61	13.42	13.23	13.04	12.84	12.65
1°C	14.38	14.19	14.00	13.82	13.63	13.44	13.26	13.07	12.88	12.70	12.51	12.32
2°C	14.01	13.82	13.64	13.46	13.28	13.10	12.92	12.73	12.55	12.37	12.19	12.01
3°C	13.65	13.47	13.29	13.12	12.94	12.76	12.59	12.41	12.23	12.05	11.88	11.70
4°C	13.31	13.13	12.96	12.79	12.61	12.44	12.27	12.10	11.92	11.75	11.58	11.40
5°C	12.97	12.81	12.64	12.47	12.30	12.13	11.96	11.80	11.63	11.46	11.29	11.12
6°C	12.66	12.49	12.33	12.16	12.00	11.83	11.67	11.51	11.34	11.18	11.01	10.85
7°C	12.35	12.19	12.03	11.87	11.71	11.55	11.39	11.23	11.07	10.91	10.75	10.59
8°C	12.05	11.90	11.74	11.58	11.43	11.27	11.11	10.96	10.80	10.65	10.49	10.33
9°C	11.77	11.62	11.46	11.31	11.16	11.01	10.85	10.70	10.55	10.39	10.24	10.09
10°C	11.50	11.35	11.20	11.05	10.90	10.75	10.60	10.45	10.30	10.15	10.00	9.86
11°C	11.24	11.09	10.94	10.80	10.65	10.51	10.36	10.21	10.07	9.92	9.78	9.63
12°C	10.98	10.84	10.70	10.56	10.41	10.27	10.13	9.99	9.84	9.70	9.56	9.41
13°C	10.74	10.60	10.46	10.32	10.18	10.04	9.90	9.77	9.63	9.49	9.35	9.21
14°C	10.51	10.37	10.24	10.10	9.96	9.83	9.69	9.55	9.42	9.28	9.14	9.01
15°C	10.29	10.15	10.02	9.88	9.75	9.62	9.48	9.35	9.22	9.08	8.95	8.82
16°C	10.07	9.94	9.81	9.68	9.55	9.42	9.29	9.15	9.02	8.89	8.76	8.63
17°C	9.86	9.74	9.61	9.48	9.35	9.22	9.10	8.97	8.84	8.71	8.58	8.45
18°C	9.67	9.54	9.41	9.29	9.16	9.04	8.91	8.79	8.66	8.54	8.41	8.28
19°C	9.47	9.35	9.23	9.11	8.98	8.86	8.74	8.61	8.49	8.37	8.24	8.12
20°C	9.29	9.17	9.05	8.93	8.81	8.69	8.57	8.45	8.33	8.20	8.08	7.96
21°C	9.11	9.00	8.88	8.76	8.64	8.52	8.40	8.28	8.17	8.05	7.93	7.81
22°C	8.94	8.83	8.71	8.59	8.48	8.36	8.25	8.13	8.01	7.90	7.78	7.67
23°C	8.78	8.66	8.55	8.44	8.32	8.21	8.09	7.98	7.87	7.75	7.64	7.52
24°C	8.62	8.51	8.40	8.28	8.17	8.06	7.95	7.84	7.72	7.61	7.50	7.39
25°C	8.47	8.36	8.25	8.14	8.03	7.92	7.81	7.70	7.59	7.48	7.37	7.26
26°C	8.32	8.21	8.10	7.99	7.89	7.78	7.67	7.56	7.45	7.35	7.24	7.13
27°C	8.17	8.07	7.96	7.86	7.75	7.64	7.54	7.43	7.33	7.22	7.11	7.01
28°C	8.04	7.93	7.83	7.72	7.62	7.51	7.41	7.30	7.20	7.10	6.99	6.89
29°C	7.90	7.80	7.69	7.59	7.49	7.39	7.28	7.18	7.08	6.98	6.87	6.77
30°C	7.77	7.67	7.57	7.47	7.36	7.26	7.16	7.06	6.96	6.86	6.76	6.66

Texas Instruments Probe Lab

Lab 13 Comparing Respiration Rates in Seeds

Problem

What factors affect the rate of respiration in peas?

Introduction

Cellular respiration is the process that releases energy by breaking down glucose and other food molecules in the presence of oxygen. All organisms, including plants and animals, oxidize glucose to obtain energy. Often, this energy is used to convert ADP and phosphate into ATP. ATP is needed to carry out basic life functions such as growth. Plants carry out both the photosynthesis and cellular respiration. Seeds cannot photosynthesize, but they do contain a food supply that can be converted to ATP. This ATP allows the embryo to grow and develop when conditions are right. However, seeds only carry out cellular respiration. What factors affect the rate of respiration in seeds?

In this lab, you will test the rate of respiration by measuring the levels of carbon dioxide under two conditions. You will explore the effects of temperature on germination and the differences between germinating and nongerminating seeds.

Skills Focus

Observe, Predict, Use a data-collection interface and sensor

Materials

- TI-83 Plus or TI-84 Plus graphing calculator
- EasyData application
- Vernier data-collection interface
- Vernier CO_2 Gas Sensor
- 250-mL respiration chamber
- 25 germinating peas
- 25 non-germinating peas
- ice cubes
- paper towel
- thermometer
- 100-mL beakers (2)
- Logger *Pro* (optional)

Safety 🔷 🔷

Always wear goggles. Wash your hands thoroughly with soap and warm water at the end of the lab.

Pre-Lab Questions

1. Recall What is the equation for cellular respiration?

_____

2. Relate Cause and Effect Explain why it is possible to compare the rates of respiration by measuring CO_2 levels.

_____

_____

_____

3. Predict Which peas do you think will have the highest rate of respiration? Explain your answer.

_____

_____

_____

_____

Procedure

1. Obtain and wear goggles.

2. Turn on the calculator. Connect the CO_2 Gas Sensor, data-collection interface, and calculator.

3. Start the EasyData application if it is not already running. Select (File) from the Main screen, and then select **New** to reset the application.

4. Measure the room's temperature using a thermometer and record it in the Temperature column of the Data Table.

5. Blot 25 germinating peas between two pieces of paper towel. Place the germinating peas into the respiration chamber. Place the shaft of the CO_2 Gas Sensor in the opening of the respiration chamber as shown in Figure 1. Wait 1 minute, and then select (Start) to begin data collection. Data will be collected for 5 minutes. When data collection has finished, a graph of CO_2 gas versus time will be displayed.

Figure 1 Completed Setup

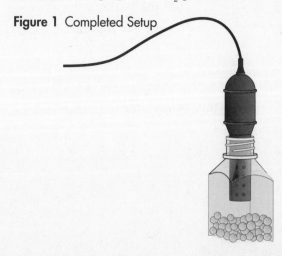

6. Remove the CO_2 Gas Sensor from the respiration chamber. Place the peas in a 100-mL beaker filled with cold water and ice cubes. Allow the peas to soak in cold water for 10 minutes. While they soak, continue to Step 7.

7. Use a notebook to fan air across the openings in the shaft of the CO_2 Gas Sensor for 1 minute. Fill the respiration chamber with water and then empty it. Thoroughly dry the inside of the respiration chamber with a paper towel.

8. Find the rate of respiration for each run, by finding the slope, a. Select ⌐Anlyz⌐, and then select **Linear Fit.** The linear-regression statistics for these two lists are displayed for the equation in the form:

$$y = ax + b$$

Record the rate of respiration in the Data Table. Select ⌐OK⌐ to view a graph of the data and the regression line.

9. Store the data from the first run so that it can be used later. Select ⌐Main⌐ to return to the Main screen. Select ⌐File⌐, and then select **Store Run.** Select ⌐OK⌐ to store your latest data and overwrite the data in Lists 3 and 4 (L3 and L4).

10. Repeat Steps 4–9, but use nongerminating peas instead of germinating peas. In Step 6, place the nongerminating peas on a dry paper towel and not in the ice bath.

11. Measure the temperature of the cold water with a thermometer and record it in the temperature column of the Data Table.

12. Remove the peas from the cold water and repeat Steps 3–6 using the cold peas.

13. Repeat steps 8 and 9.

14. Clean up as directed by your teacher.

Collect Data

15. Graph all three runs of data on a single graph. To do this, select ⌐Adv⌐ and then select **L2, L3 and L4 vs L1.** All three runs should now be displayed on the same graph. Each point of the germinating peas is plotted with a small plus sign, each point of the nongerminating peas is plotted with a square, and each point of the cold peas is plotted without a marker. Examine the graph closely to help you answer the Analyze and Conclude questions. When finished with the graph, select ⌐Main⌐ to return to the Main screen.

Data Table		
Peas	Temperature (°C)	Rate of Respiration (ppm/min)
Germinating		
Nongerminating		
Germinating		

Analyze and Conclude

1. Interpret Graphs What evidence is there that cellular respiration has occurred?

_____

_____

2. Interpret Data What is the effect of germination on the rate of cellular respiration in peas?

_____

_____

_____

_____

3. Interpret Data What is the effect of temperature on the rate of cellular respiration in peas?

_____

_____

_____

4. Apply Concepts Why do germinating peas undergo cellular respiration?

_____

_____

_____

5. Predict How would the rate of cellular respiration change if you add water to the dry peas, keeping them at room temperature?

_____

_____

_____

Extend Your Inquiry

Is respiration equal between different organisms? Design an experiment to test the rate of respiration in different organisms. You could test germination of different types of seeds or use a small animal such as insects or earthworms. **IMPORTANT:** If your experimental design involves the use of animals, be sure to handle them with care. Avoid harming them or youself.

Lab 14 Transpiration Rates

Problem

What factors affect the rate of transpiration in plants?

Introduction

Water transport is an essential process in land plants. Water is obtained
by roots at the bottom of the plant but is used by leaves at the top.
Efficiently transporting water between the roots and leaves is needed
for the plant's survival. In leaves, the walls of mesophyll cells are kept
moist so that gases can enter and leave the cells easily. As a result, water
continuously evaporates from the surface of leaves in a process known
as transpiration. As each water molecule evaporates, cohesion pulls
more water molecules out from the xylem vessels in the leaf. This pull is
transmitted all the way down from the leaves to the root in a continuous
water column, allowing the efficient movement of water from the roots
up to the leaves. Any factor that affects a plant's transpiration rate affects
water transport throughout the plant. This lab will measure transpiration
rates under different conditions of light, humidity, temperature, and air
movement. The data will be collected by measuring pressure changes as
the plant takes up water into the stem.

Skills Focus

Observe, Predict, Use a Calculator, Data-Collection Interface and Sensor

Materials

- TI-83 Plus or TI-84 Plus graphing
  calculator
- EasyData application
- Vernier data-collection interface
- Vernier Gas Pressure Sensor, with
  provided plastic tube
- 3 utility clamps
- ring stand
- tomato seedlings
- plastic tube clamp
- dropper or Beral pipette
- scalpel

- thermometer
- metric ruler
- masking tape
- 100-watt light source
- plastic gallon-sized bag with
  twist tie
- heater, small electric
- fan with slow speed
- plant mister
- plastic syringe
- shallow bowl, with hard walls

Safety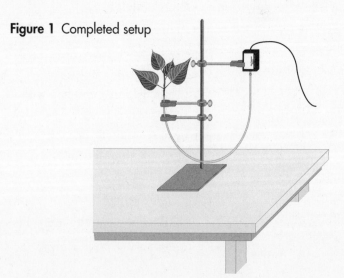

Always wear goggles. Do not direct the point of the scalpel toward yourself or others. Use the scalpel only as instructed. Wash your hands thoroughly with soap and warm water at the end of the lab.

Pre-Lab Questions

1. **Recall** Which specialized cells regulate the movement of gases into and out of leaf tissue?

_____

2. **Apply Concepts** In Step 4 of the lab, why is it important to submerge the plant under water?

_____

_____

_____

3. **Predict** Which environmental condition do you think will have the greatest effect on the rate of transpiration?

_____

_____

_____

Procedure

Part A: Setting Up the Control

1. Position the ring stand, utility clamps, and Gas Pressure Sensor as shown in Figure 1. Do not place the plastic tube or the plant in the apparatus yet.

Figure 1 Completed setup

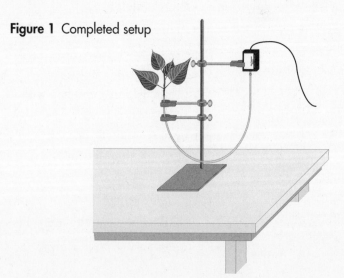

2. Place the tip of the plastic syringe into the end of the plastic tube without the white plastic connector. Submerge the other end of the tube into water. Use the syringe to draw water up into the tube until it is full. Check the tube closely for air bubbles. If you see any, you may try to expel them by tapping the tube. If this does not work, start the process of filling the tube over again.

3. Slip a plastic tube clamp on the tube as shown in Figure 2. Bend the tube into a U shape with both ends up, being careful to hold both ends at equal heights. If one end is lower than the other, water will drip out. With the tube full of water, remove the syringe.

Figure 2 Tube clamp on tube

4. Select a plant that has a stem about the same diameter as the opening of the tube. Use a scalpel to carefully cut the plant 1 inch above the soil. Submerge the plant under water in a shallow bowl. Against the bottom of the bowl, carefully make a cut at a 45° angle near the base of the stem.

5. Raise the end of the tube with the white plastic connector until water begins to drip out of the other end. Carefully push the cut stem of the plant into the end of the tube where the water is dripping out. Be careful not to allow any air bubbles to form between the cut portion of the stem and the water in the tube. Push the plant at least 1 centimeter into the tube. Be careful not to damage the plant. Shut the tube clamp as tightly as possible, as shown in Figure 3.

Figure 3 Clamping plant in tube

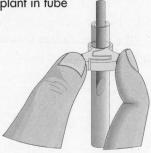

6. When the tube clamp is shut tight, invert your plant cutting to check for any leaks. If water does leak out, turn the plant right side up and try tightening the clamp further.

   IMPORTANT: Be sure the tube is completely filled with water. The water column must be flush with the stem. There should be no air visible at the base of the stem. If water moves down the tube away from the stem after it has been inserted, check for a leak in the system.

7. Connect the opposite end of the plastic tube to the sensor valve.
   CAUTION: Do not allow water to enter the valve of the Gas Pressure Sensor.

8. Secure the plant in an upright position with the utility clamps as shown in Figure 1. It should be positioned so that the cut stem is about 8 cm below the water level at the other end of the tube, as shown in Figure 1. On the end of the tube without the plant, place a mark at the water level to allow you to refill the tube to the starting level in Step 18.

9. Review your setup, and check it carefully against the diagram in Figure 1. The water column must be continuous for the experiment to work. If any air bubbles have formed, you may need to start over.

10. Place your plant setup in an area where the wind, humidity, and temperature are reasonably constant. This location will be your control setup. If you move your plant, be sure to hold the ends of the tubing steady, so they do not move up or down relative to each other.

11. Allow the system 5 minutes to adjust to the environment. While the system is adjusting, set up your data-collection interface.

12. Turn on your calculator. Connect the Gas Pressure Sensor, data-collection interface, and calculator. Start the EasyData application if it is not already running. Select from the Main screen, and then select **New** to reset the application.

13. Check the base of the plant stem in the water tube to make sure that no air bubbles or air pockets have formed that will prevent the plant from taking up water. If an air pocket has formed, refit the plant in the tube before initiating data collection in Step 14.

14. After the plant has equilibrated for 5 minutes, select (Start) to begin data collection. Data will be collected for 15 minutes. During that time, if it is necessary to quit data collection early, you can select (Stop) to end data collection.

15. When data collection has stopped, perform a linear regression to calculate the rate of transpiration.

    a. Select (Anlyz), and then select **Linear Fit**.

    b. The linear-regression statistics for these two lists are displayed for the equation in the form:

    $$y = ax + b$$

    c. Enter the absolute value of the slope, a, as the rate of transpiration in Data Table 1.

    d. Select (OK) to view a graph of the data and the regression line.

    e. Select (Main) to return to the Main screen.

Part B: Design Your Own Experiment

16. As a class, you will test the effect of a variety of environmental factors on transpiration. Each group will be assigned by your teacher to test the effect of one of the following environmental factors: light intensity, wind, humidity, temperature.

17. Design an experiment to simulate an increase or decrease in the environmental factor you are assigned. Check your procedure with your teacher and obtain the materials needed for the experiment.

18. Refill the water level in the tube to the same level marked in Step 8. Be sure there are no air bubbles, and then begin your data collections. Record your values in Data Table 1.

Data Table 1					
Test	Slope (kPa/min)	Mass (g)	Surface area (cm²)	Rate/Area (kPa/min/cm²)	Adjusted Rate (kPa/min/cm²)
Simulated Environmental Condition					
Control					

Part C: Determine the Rate of Transpiration

19. Cut all the leaves (not stems) off your plant. Put all of the leaves on a balance and measure the mass. Record the value in Data Table 1.

20. To find the surface area, cut out a section of leaf 5 cm × 5 cm. Determine the mass for this leaf section and divide by 25 cm^2 to find the mass of l cm^2 of leaf. Divide the total mass of the leaves by the mass of l cm^2 to find the total leaf surface area. Record the calculated surface area in Data Table 1.

21. Calculate the rate of transpiration. To do this, divide the slope by the surface area for each plant. These rate values can be expressed as kPa/min/cm^2. Record the rate of transpiration in Data Table 1.

22. Find the adjusted rate for your simulated condition by subtracting the rate of transpiration of the control from the rate of transpiration of the simulated condition. Record this adjusted rate in the last column of Data Table 1.

23. Share the adjusted rate for your environmental condition with the class. Record the class results in Data Table 2 for each of the tested environmental conditions. If a condition was tested by more than one group, take the average of the values and record it in Data Table 2.

Data Table 2	
Class Data	
Simulated Environmental Condition	Adjusted Rate (kPa/min/cm^2)
Light	
Humidity	
Wind	
Temperature	

24. **Graph** Make a bar graph that shows the effect of different environmental conditions on the transpiration in plant cuttings. Use the data in Data Table 2 to plot the adjusted rate for each test on the y-axis and the environmental condition on the x-axis.

Analyze and Conclude

1. **Interpret Graphs** Which simulated environmental condition resulted in the greatest rate of water loss? Explain why this factor might increase water loss when compared to the others.

_____

_____

_____

_____

2. **Draw Conclusion** Under which of the tested environmental conditions would you expect to see more closed stomata? Use your data to explain your answer.

_____

_____

_____

_____

3. **Applying Concepts** Why is it necessary to determine the surface area of the leaves in order to calculate the rate of transpiration?

_____

_____

_____

_____

4. **Connect Concepts** Scientists sometimes measure transpiration rates and use them to figure out how much photosynthesis is occurring. Explain how this works.

_____

_____

_____

_____

Extend Your Inquiry

Test the rate of transpiration in different types of plants. Set up a similar experiment, but use two or three different types of plants.

Lab 15 **Human Respiration**

Problem

What affects the rate of respiration in humans?

Introduction

Through the process of breathing, your body obtains oxygen, releases carbon dioxide, and regulates the blood's pH level. The process of taking in air is known as *inhalation,* while the process of blowing out air is called *exhalation.* The rate of respiration depends on the levels of oxygen and carbon dioxide in your blood.

In this lab, you will monitor the respiratory patterns of one member of your group under different conditions. A respiration belt will be strapped around the test subject and connected to a calculator-interfaced Gas Pressure Sensor. Each respiratory cycle, one inhalation and one exhalation, will be recorded by the interface, allowing you to calculate and compare respiratory rates under different conditions.

Skills Focus

Use a Calculator, Data-Collection Interface and Sensors, Analyze Data, Infer

Materials

- TI-83 Plus or TI-84 Plus graphing calculator
- EasyData application
- Vernier data-collection interface
- Vernier Gas Pressure Sensor
- Vernier Respiration Monitor Belt
- respiration bag

Safety

If you are prone to dizziness or nausea, do not volunteer to be the test subject. If the person being tested experiences dizziness, nausea, or a headache during testing, stop data collection and inform your teacher.

Pre-Lab Questions

1. **Use Models** In Part C of this lab, what type of situation does rebreathing exhaled air simulate?

_____

_____

_____

2. **Predict** Under which condition will the test subject's rate of respiration be greater?

_____

_____

_____

Procedure

Part A: Set Up the Experiment

1. Turn on the calculator. Connect the Gas Pressure Sensor, data-collection interface, and calculator.

2. Set up EasyData for data collection with a Respiration Monitor Belt.

   a. Start the EasyData application, if it is not already running.

   b. Select ⌐File⌐ from the Main screen, and then select **New** to reset the application.

   c. Select ⌐Setup⌐ from the Main screen, and then select **Other Sensors...**

   d. With **CH1** highlighted, select ⌐Next⌐.

   e. Select **Respiration**, and then select ⌐Next⌐ to return to the Main screen.

   f. Select ⌐Setup⌐ again, and then select **Rate...**

   g. Select ⌐Edit⌐ from the Rate Mode Settings screen.

   h. Enter **10** seconds as the time between samples and select ⌐Next⌐.

   i. Enter **18** as the number of samples and select ⌐Next⌐. (Data will be collected for 3 minutes.)

   j. Select ⌐OK⌐ to return to the Main screen.

3. Select a member of your lab group as the test subject. Wrap the Respiration Monitor Belt snugly around the subject's rib cage. Press the hook and loop fastener strips together at the back. Position the belt so the air bladder is resting over the diaphragm as shown in Figure 1.

Figure 1 Completed setup

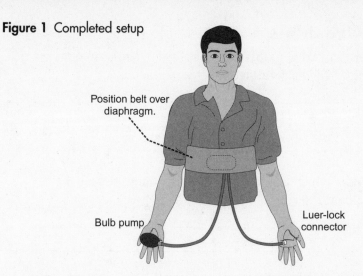

Position belt over diaphragm.

Bulb pump

Luer-lock connector

4. Attach the Respiration Monitor Belt to the Gas Pressure Sensor. There are two rubber tubes connected to the bladder. One tube has a white Luer-lock connector at the end, and the other tube has a bulb pump attached. Connect the Luer-lock connector to the stem on the Gas Pressure Sensor with a gentle half turn.

5. Have the test subject sit upright in a chair. Close the shut-off screw of the bulb pump by turning it clockwise as far as it will go. Pump air into the bladder by squeezing the bulb pump. Fill the bladder as full as possible without making it uncomfortable for the test subject.

Part B: Holding of Breath

6. Select ⌈Start⌉ on the calculator to begin data collection. Data will be collected for 3 minutes. For the first 30 seconds, the test subject should breathe normally. When 30 seconds have passed, the test subject should take a deep breath and hold it as long as he or she can without experiencing extreme discomfort or concern. After the subject releases his or her breath, instruct him or her to breathe normally for the remainder of data collection.

7. When data collection has finished, a graph of respiration versus time will be displayed. As you move the cursor right or left, the time (X) and respiration rate (Y) values of each data point are displayed above the graph.

8. Determine the respiration rate before and after the test subject's breath was held and record these rates in Data Table 1. To do this:

   a. Select ⌈Anlyz⌉, and then select **Statistics...**

   b. Set your left boundary by using ⓓ and ⓒ. Move the cursor to three points before the subject's breath was held and select ⌈OK⌉.

   c. Set your right boundary by using ⓓ and ⓒ. Move the cursor to the point just before the subject's breath was held and select ⌈OK⌉. The application will now calculate and display the statistics for the data between the two boundaries.

d. Record the mean value in Data Table 1 as the respiration rate before the subject held his or her breath. Then select ⌜OK⌝ to return to the Graph screen.

e. Select ⌜Anlyz⌝ and then **Statistics...** to determine the mean respiration rate after the subject's breath was held.

f. Set your left boundary by using the cursor keys. Move the cursor to the point just after the subject's breath was released and select ⌜OK⌝.

g. Set your right boundary by using the cursor keys. Move the cursor three points beyond where the subject's breath was released and select ⌜OK⌝. The application will now calculate and display the statistics for the data between the two boundaries.

h. Record the mean value in Data Table 1 as the respiration rate after the subject held his or her breath.

i. Select ⌜OK⌝, and then select ⌜Main⌝ to return to the Main screen.

Part C: Breathing Exhaled Air

9. Prepare data-collection interface and calculator for a new data collection. **IMPORTANT**: Do not select Respiration for this part of the experiment.

a. Select ⌜File⌝ from the Main screen, and then select **New** to reset the application.

b. Select ⌜Setup⌝ from the Main screen, then select **Time Graph...**

c. Select ⌜Edit⌝ on the Time Graph Settings screen.

d. Enter **0.5** seconds as the time between samples and select ⌜Next⌝.

e. Enter **120** as the number of samples and select ⌜Next⌝. (Data will be collected for 1 minute.)

f. Select ⌜OK⌝ to return to the Main screen.

10. Obtain a respiration bag from your teacher. The subject will breathe in and out of the bag through the attached cup during this portion of the experiment.

11. The subject will breathe in and out of the respiration bag for a period of 6 minutes. During the 6-minute period, three series of data will be collected. Use a watch or clock to make sure that data collection is performed at the proper times. **IMPORTANT**: If the test subject experiences dizziness, nausea, or a headache during data collection, testing should be stopped immediately, and the subject should remove the respiration bag so he or she can breathe normally.

12. Fill the respiration bag with air and have the subject hold the breathing cup snugly to his or her face so that outside air does not get in.

13. Have the subject begin breathing into and out of the bag through the attached cup. Check to make sure that he or she is able to breathe easily both in and out of the respiration bag. The subject must breathe from the bag for the entire 6 minutes. Select ⌜Start⌝ to begin collecting data. Data will be collected for 1 minute. During that time, look ahead to Steps 14 and 15 and prepare to follow the instructions.

14. When 1 minute has elapsed, data collection will stop.

   a. One member should perform Step 15 **immediately**.

   b. Another member of your lab group must continue to keep track of the total elapsed time.

   c. The subject must continue breathing into the bag for the entire 6-minute period.

15. Calculate the relative amplitude of the respiration waves for the data collection. You must complete this step within 1 minute.

   a. Select ⌈Anlyz⌉, and then select **Statistics...**

   b. With the cursor to the left-most boundary of the displayed curve, select ⌈OK⌉ to set it.

   c. The cursor will move to the right side of the graph. Ensure that the cursor is on the last data point and select ⌈OK⌉ to set the right boundary. The program will now calculate and display the statistics for the data between the two boundaries.

   d. Subtract the minimum pressure from the maximum pressure to calculate the amplitude.

   e. Record the calculated amplitude for the time interval in Data Table 2.

   f. Select ⌈OK⌉ to return to the Graph screen, then ⌈Main⌉ to return to the Main screen.

16. After a total of 2.5 minutes have elapsed, with the subject still breathing into the bag, select ⌈Start⌉, then ⌈OK⌉ to overwrite the previous data and begin collecting respiration data from the subject. Repeat Steps 14–15, and record the amplitude value in Data Table 2 under "2.5 to 3.5 Minutes."

17. After a total of 5 minutes have elapsed, with the subject still breathing into the bag, select ⌈Start⌉, then ⌈OK⌉ to overwrite the previous data and begin collecting respiration data from the subject. Repeat Steps 14–15, and record the amplitude value in Data Table 2 for "5 to 6 Minutes."

Data Table 1	
Holding of Breath: Respiration Rates	
Before Holding Breath	After Holding Breath
_____ breaths / minute	_____ breaths / minute

Data Table 2		
Breathing Exhaled Air: Amplitudes of Respiration Waves		
0 to 1 Minute	2.5 to 3.5 Minutes	5 to 6 Minutes
_____ kPa	_____ kPa	_____ kPa

Analyze and Conclude

1. **Analyze Data** Did the rate of respiration of the test subject change after holding his or her breath? If so, describe how it changed.

_____

_____

_____

_____

2. **Draw Conclusions** What is the significance of an increase in the amplitude and frequency of the waveform while the test subject was breathing into the bag?

_____

_____

_____

_____

3. **Interpret Data** How did the respiratory waveforms change while the test subject was breathing exhaled air from the bag? How would you interpret this result?

_____

_____

_____

_____

4. **Apply Concepts** Based on your results, how does carbon dioxide affect your breathing?

_____

_____

_____

_____

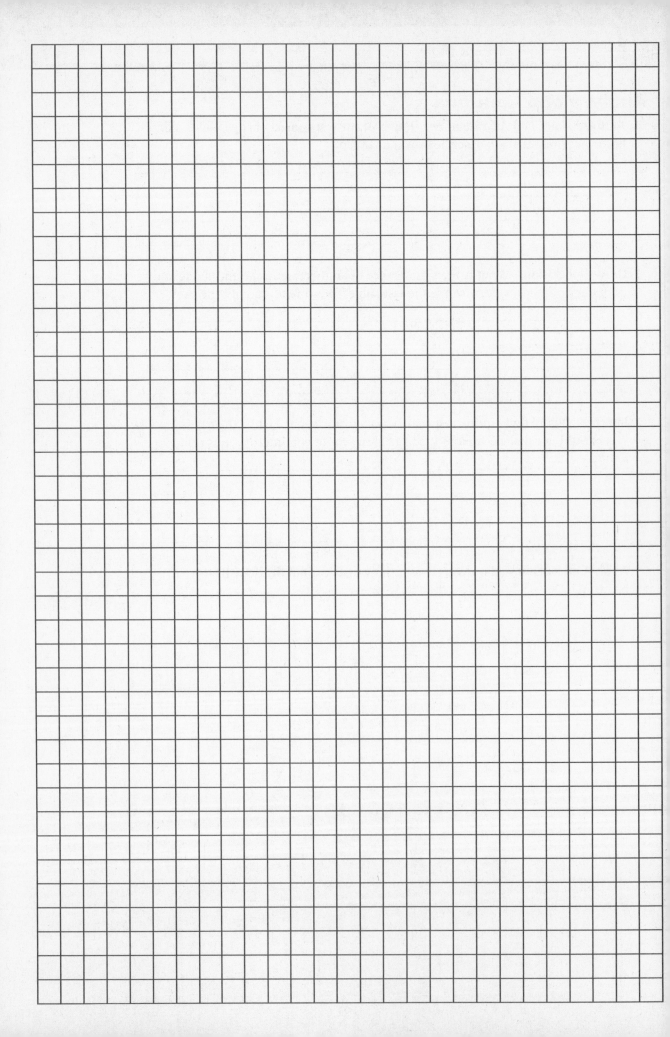

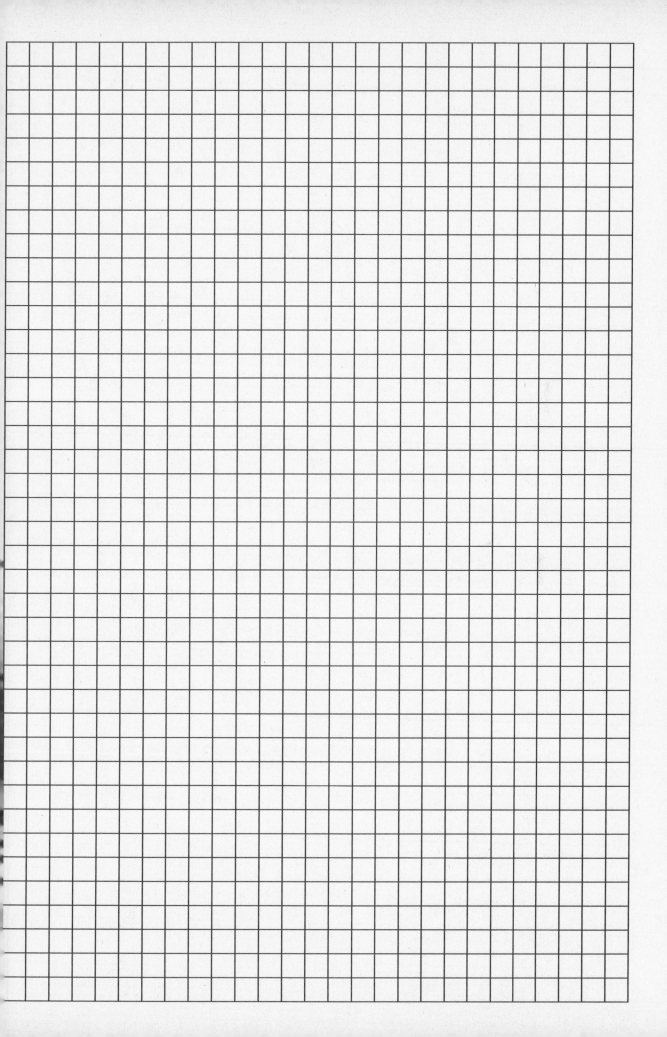

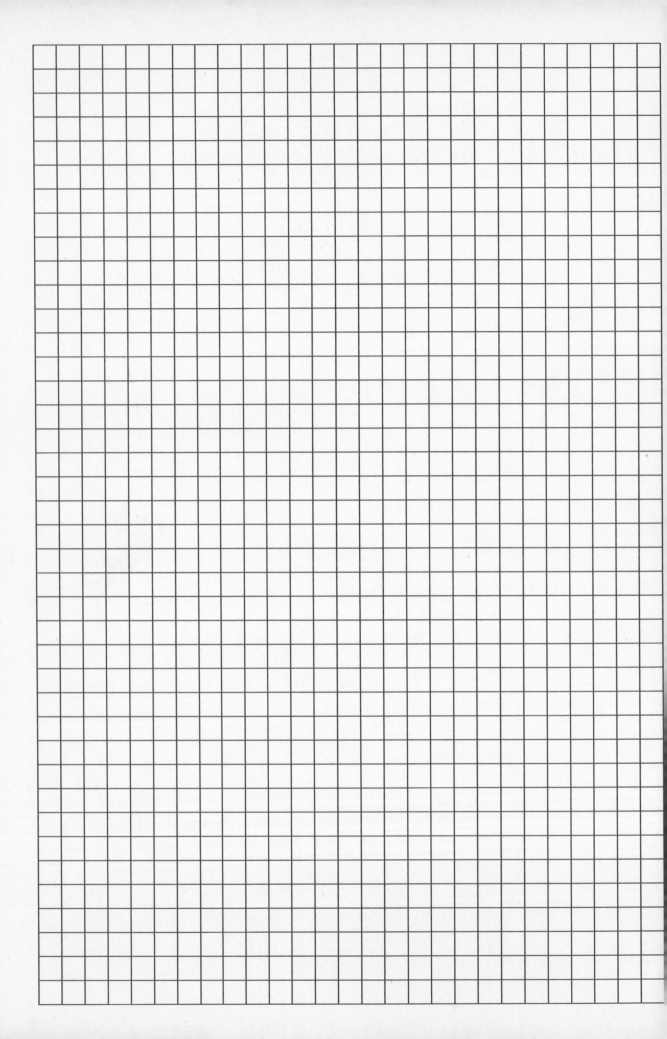